José Luis Cort · Pablo Abaunza

The Bluefin Tuna Fishery in the Bay of Biscay

Its Relationship with the Crisis of Catches of Large Specimens in the East Atlantic Fisheries from the 1960s

José Luis Cort
Centro Oceanográfico de Santander
Spanish Institute of Oceanography
Santander, Cantabria, Spain

Pablo Abaunza
Spanish Institute of Oceanography
Madrid, Spain

https://doi.org/10.1007/978-3-030-11545-6

Atlantic bluefin tuna (Courtesy of *Grup Balfegó*)

Acknowledgements

Thanks are extended to Bernardo Pérez and Pablo Barquín (*IEO*, Santander) for their help in preparing the graphics and manuscripsts of the present study; to Juan J. Navarro (*Grup Balfegó*) for permission to use the images provided; to Cesar González-Pola (*IEO*) for the design of the Fig. 2.2 showing the currents in the Bay of Biscay; to the Institute of Marine Reseach in Norway (IMR) for providing the photograph of Fig. 5.11; to Øyvind and Magnus Tangen for providing figures presented at the Santander Symposium; to *Taylor and Francis Group* for allowing the publication of several figures and tables from two publications; to Victor Díaz del Río, oceanographer geologist of the IEO, for the description of the Strait of Gibraltar in the times of Neanderthal man; to Francisco Abascal (*IEO*) for facilitating the geolocation of Fig. 7.3; to Carlos Palma, ICCAT biostatistician, always willing to clarify and facilitate information, and for facilitating the Figs. 2.4 and 8.3; and to Antonio Di Natale (Genoa Aquarium), for the revision of the manuscript.

Contents

Abstract

A brief description is made of the Atlantic bluefin tuna, *Thunnus thynnus* (L.), from the point of view of its biology, geographic distribution and population dynamic. With the Strait of Gibraltar traps as a reference and providing a vision from ancient times to the present, scientific activities on this fishing gear are reviewed in an attempt to explain the reasons behind the crisis of the catches in this fishery, which began at the beginning of the 1960s and which has been extended until recent times. In order to do so and based on a recent publication on the issue, the Bay of Biscay fishery is presented as a fishery with a long tradition on which numerous scientific studies have been carried out and whose data are of great importance in the ABFT assessment group of SCRS (ICCAT's scientific committee). Traditionally, the catches of this fishery have been made up of juvenile specimens (<40 kg) making up more than 97% in number of fishes according to a 62-year series studied (1949–2010). From an analysis of the population, it has been shown that the impact of massive fishing of juveniles in this fishery between 1949 and 1962, together with that of juveniles in Morocco (the Atlantic part) possibly caused the decline of the traditional traps fishery of spawners in the Strait of Gibraltar and the collapse of the spawner fisheries in the north of Europe from the 1960s. This situation continued until 2007, the year from which a Pluri-Annual Recovery Plan (PARP) was adopted by ICCAT for the eastern stock. Since the PARP was implemented, juvenile catches have disappeared from almost all the fisheries, a scenario that coincides with a considerable recovery of the spawning stock biomass in the easter Atlantic and Mediterranean, according to the results of the latest SCRS assessments.

Keywords Atlantic Bluefin Tuna · Traps · Strait of Gibraltar · Bay of Biscay · Juveniles Fishing

Chapter 1
Introduction

Atlantic bluefin tuna, *Thunnus thynnus* (L.), from the Atlantic and Mediterranean, hereinafter referred to as ABFT, has undergone various crises in recent times owing to overfishing, which is why the *International Commission for the Conservation of Atlantic Tunas* (ICCAT) recently adopted highly restrictive conservation measures for the protection of the resource. The ICCAT Standing Committee on Research and Statistics (SCRS) carries out periodic assessments of resources to examine the impact of such conservation measures on the population.

Since the middle of the 20th century fishing pressure on ABFT resources has been growing constantly, causing increasing damage to stocks. New fisheries have been created and fishing effort has increased throughout the Atlantic and Mediterranean to the extent that on two occasions, one in 1993 (Safina 1993) and the other in 2009 (Fromentin et al. 2014) the species' inclusion under Appendix 1 of the *Convention of International Trade in Endangered Species of Wild Fauna and Flora* (CITES) was proposed. Were it to be included it would mean the prohibition of international trade in ABFT. This did not happen on those two occasions, but it was a close call. ICCAT's intervention was crucial on both occasions in redirecting the situation and returning the resource to a situation of apparent sustainability.

Among the most relevant events in recent decades the following can be highlighted:

- Overfishing in the eastern Atlantic in the 1960s (when ICCAT had not yet been founded), which led to a permanent crisis in the traps fishery of the Strait of Gibraltar and the collapse of the northern European fisheries (Cort and Abaunza 2015). From the 1970s purse seine fishing developed rapidly in the Mediterranean, which became the main fishing gear for the catch of ABFT in this sea (ICCAT 2010). Meanwhile, the spawner fisheries of the eastern Atlantic remained at very low levels from the beginning of the 1960s and in 1974 led ICCAT to adopt the first conservation measure (ICCAT 1974), which established a minimum catchable length of age 1 fishes ($W = 6.4$ kg). This measure was not implemented in any eastern stock fisheries until the beginning of the 2000s, which for decades resulted in the illegal catch of millions fishes below the minimum length.

J. L. Cort and P. Abaunza, *The Bluefin Tuna Fishery in the Bay of Biscay*,
SpringerBriefs in Biology, https://doi.org/10.1007/978-3-030-11545-6_1

- At the beginning of the 1980s a second ABFT crisis arose as a result of overfishing, this time in the fisheries of the western stock (a definition of a stock is provided later), a crisis that led to ICCAT's adoption of the first TACs (Total Allowable Catches) for this species, which have been set between 2000 and 2500 t/year since 1982 (Fromentin and Powers 2005). Later, following the recent disappearance of traditional fisheries such as the Sicilian traps, the continued proliferation of purse seine fleets in the Mediterranean and illegal fishing by vessels with flags of convenience had taken ABFT catches to an official figure of 50,000 t in 1995 (ICCAT 2014; Fromentin et al. 2014), in 1998 ICCAT adopted the first TACs for the eastern stock fisheries (32,000 t/year), though the quantity recommended by the scientists at that time was between 15,000 and 25,000 t/year. Nevertheless, following the adoption of this conservation measure there was a complete lack of monitoring among the fleets and the measure was only implemented in the traps.
- Lastly, the Mediterranean overfishing in the 2000s coincided with the start of purse seine fishing for fattening on farms, an activity that brought about the plundering of the species in the Mediterranean (WWF 2008). According to estimations made by the SCRS, the real catch between 1998 and 2006 was between 50,000–61,000 t/year, though the TAC remained at 32,000 t (Fromentin et al. 2014). The scientific reports of 2008 and 2010 pointed to a fall in the spawning biomass of up to 80% below that of the maximum sustainable yield (ICCAT 2012; ICCAT 2014). In view of all this, in 2007 ICCAT adopted a Pluri-Annual Recovery Plan (PARP) throughout all the fisheries of the eastern Atlantic and Mediterranean (ICCAT 2006) involving a very low TAC (for example, 12,900 t in 2012) among other measures such as limiting the minimum weight at catch to 30 kg (ages 1–4), all under the management of the international commission. There were, however, some exceptions to the minimum weight, such as in the Bay of Biscay, Adriatic Sea and the artisanal coastal Mediterranean fishery, where it is just 8 kg (ICCAT 2007).

The PARP has important consequences:

(i) It has led to a considerable reduction in the TAC, which has in turn led to a large fall in the number of fishing vessels in the Mediterranean; (ii) the increase in the minimum size of ABFT (from 10 to 30 kg) has brought about the disappearance of most of the juvenile fisheries; and (iii) it has brought with it strict monitoring of the landings of this species. All of this has meant that in the ten years that the PARP has been in force the spawning stock biomass of ABFT has increased very significantly, something to which the high recruitments in some years may have contributed as a result of favorable environmental conditions (Piccinetti et al. 2013). In the most recent assessments (ICCAT 2012, 2014, 2017) and from different fishing indicators available to the assessment group [the traps in Sardinia (Addis et al. 2012); the Moroccan traps of the Atlantic (Abid et al. 2017); the larvae of the Balearic Sea (Álvarez-Berastegui et al. 2018); the Spanish purse seiners in the Balearic Sea (Gordoa 2014, 2017); the Portuguese traps in the Strait of Gibraltar (Lino et al. 2018); the Japanese long liners in the Atlantic (Kimoto and Itoh 2017); the aerial surveys in the Gulf of Lion (Rouyer et al. 2018); and the Tunisian purse seiners in the central Mediterranean (Zarrad and Missaoui 2017)] a very considerable recovery of resources has been recorded, which has given rise to a continual increase in quotas (TAC = 36,000 t by 2020) which may bring the PARP to an end.

Since the very first fishing season following the implementation of the PARP (2008), a minimum of 840,000 juvenile specimens (<30 kg) were no longer being caught each year in the central and western Mediterranean (Cort and Martinez 2010). This figure is mainly based on the catches of the 83 vessels of Italy, France and Spain (>40 m) registered at ICCAT, which caught around 100–150 t/year of juveniles according to estimates made by the SCRS (ICCAT 2008). Most of these juveniles have now joined the spawning stock and this is possibly one of the reasons behind the continuing rise of the abundance indices of spawners since then. A simulation made by Belda and Cort (2011) in a scenario in which there were no juvenile catches (<30 kg) by the EU fleet reveals the same trend and quantity of biomass as that reached in the assessment of the ABFT group in 2017 (ICCAT 2017).

Since the adoption and implementation of the PARP the situation has been completely turned around, such that now, under the monitoring of fisheries by member states, the resources of this species are very much recovered (ICCAT 2014, 2017). Moreover, scientific activities on ABFT have since multiplied (Di Natale et al. 2018).

The present study focuses on one of the events described above: overfishing in the 1950s, which was what brought about the disappearance of most of the traps of the Strait of Gibraltar and led to the collapse of the northern European fisheries from the 1960s; the answers provided by science concerning these facts; how the catches of the juvenile fisheries affected this crisis; and how the crisis in catches has now been overcome.

To deal with this matter, and considering the traps as a central theme of the study, the following chapters are described:

- A brief review of the ABFT, which includes some characteristics such as the description of the species, its habitat, stocks, migrations, growth, reproduction, fishing, etc.
- The catch of the species, taking into account the traps of the Strait of Gibraltar, making a tour from their ancient history to the present. This section includes accounts based on archaeological research in the Strait of Gibraltar.
- The trap fisheries from a historical point of view, though given the immense bibliography on this subject and the numerous scientists involved in it, scientific activities related to this fishing gear by Spanish scientists from the twentieth century are cited.
- ABFT fisheries in the eastern Atlantic, the interaction among them and associated research. Special attention is paid to the Bay of Biscay fishery, both from the historical point of view and from fisheries research, as this is now one of the most closely studied ABFT fisheries.
- The latest assessments of ABFT resources carried out by the SCRS group of experts, the most important scientific reference regarding the situation of the stocks of this species.

References

Abid N, Malouli M, Ben Mhamed A (2017) Standardized CPUE of bluefin tuna, *Thunnus thynnus*, caught by Moroccan Atlantic traps for the period 1986–2016. ICCAT SCRS/2017/038, 7 p

Addis P, Secci M, Locci I, Cau A, Sabatini A (2012) Analysis of Atlantic bluefin tuna catches from the last *Tonnara* in the Mediterranean Sea: 1993–2010. Fish Res 127(128):133–141. https://doi.org/10.1016/j.fishres.2012.05.010

Álvarez-Berastegui D, Ingram GW Jr, Reglero P, Ferrà C, Alemany F (2018) Changes of bluefin tuna (*Thunnus thynnus*) larvae fishing methods over time in the western Mediterranean, calibration and larval indices updating. Col Vol Sci Pap ICCAT 74(6):2772–2783

Belda E, Cort JL (2011) Simulation of biomass trends of eastern bluefin tuna (*Thunnus thynnus*) stock under current management regulations. Col Vol Sci Pap ICCAT 66(2):989–994

Cort JL, Abaunza P (2015) The fall of tuna traps and collapse of the Atlantic bluefin tuna, *Thunnus thynnus* (L.), fisheries of Northern Europe in the 1960s. Rev Fish Sci Aquac 23:4, 346–373. http://dx.doi.org/10.1080/23308249.2015.1079166

Cort JL, Martínez D (2010) Posibles efectos del Plan de Recuperación de atún rojo (*Thunnus thynnus*) en algunas pesquerías españolas. Col Vol Sci Pap ICCAT 65:868–874

Di Natale A, Lino P, López González JA, Neves dos Santos M, Pagá García A, Piccinetti C, Tensek S (2018) Unusual presence of small bluefin tuna YOY in the Atlantic Ocean and in other areas. Col Vol Sci Pap ICCAT 73(6):3510–3514

Fromentin JM, Powers J (2005) Atlantic bluefin tuna: population dynamics, ecology, fisheries and management. Fish 6:281–306

Fromentin JM, Bonhommeau S, Arrizabalaga H, Kell LT (2014) The spectre of uncertainty in management of exploited fish stocks: the illustrative case of Atlantic bluefin tuna. Mar Policy 47:8–14

Gordoa A (2014) Catch rates and catch size structure of the Balfegó purse seine fleet in Balearic waters from 2000 to 2014; two years of size frequency distribution based on video techniques. Col Vol Sci Pap ICCAT 71(4):1803–1812

Gordoa A (2017) Updated bluefin tuna CPUE and catch structure from Balfegó purse seine fleet in Balearic waters from 2000 to 2016. Col Vol Sci Pap ICCAT 73(6):2020–2025

ICCAT (1974) Recomendación 74-1. Talla límite y mortalidad por pesca del atún rojo. https://www.iccat.int/Documents/Recs/compendiopdf-s/1974-01-s.pdf

ICCAT (2006) Informe del período bienal 2006–2007, 1ª parte (2006), vol 1. http://www.iccat.int/Documents/BienRep/REP_ES_06-07_I_1.pdf

ICCAT (2007) Report for biennial period, 2006–07. Part I (2006), vol 2, 240 p. https://www.iccat.int/Documents/BienRep/REP_EN_06-07_I_2.pdf

ICCAT (2008) Report for biennial period, 2006–07. Part II (2007), vol. 1, 276 p

ICCAT (2010) ICCAT manual. Description of species. Chapter 2; 2.1.5 Atlantic bluefin tuna, vol 99. Madrid, ICCAT, pp 93–111. http://iccat.int/Documents/SCRS/Manual/CH2/2_1_5_BFT_ENG.pdf

ICCAT (2012) Report of the 2012 Atlantic bluefin tuna stock assessement session. Madrid, Spain, 124 p, 4–11 September 2012. http://www.iccat.int/Documents/Meetings/Docs/2012_BFT_ASSESS.pdf

ICCAT (2014) Report of 2014 Atlantic bluefin tuna stock assessment session. Madrid, Spain, 178 p, 20–28 July 2014. http://iccat.int/Documents/Meetings/Docs/2014_BFT_ASSESS-ENG.pdf

ICCAT (2017) Report of the 2017 ICCAT bluefin stock assessment meeting. Madrid, Spain, 106 p, 22–27 July 2017. http://iccat.int/Documents/Meetings/Docs/2017_BFT_ASS_REP_ENG.pdf

Kimoto A, Itoh T (2017) Simple update of the standardized bluefin tuna CPUE of Japanese longline fishery in the Atlantic up to 2016 fishing year. Col Vol Sci Pap ICCAT 73(6):1957–1976

Lino P, Rosa D, Coelho R (2018) Update pn the bluefin tuna catches from the tuna trap fishery off southern Portugal (NE Atlantic) between 1998 and 2016, with a preliminary CPUE standardization. Col Vol Sci Pap ICCAT 74(6):2719–2733

Piccinetti C, Di Natale A, Arena P (2013) Eastern bluefin tuna (*Thunnus thynnus*, L.) reproduction and reproductive areas and season. Col Vol Sci Pap ICCAT 69(2):891–912

Rouyer T, Brisset B, Bonhommeau S, Fromentin JM (2018) Update of the abundance index for juvenile fish derived from aerial surveys of bluefin tuna in the western Mediterranean Sea. Col Vol Sci Pap ICCAT 74(6):2887–2902

Safina C (1993) Bluefin tuna in the West Atlantic: negligent management and the making of an endangered species. Originally published in Conservation Biology, vol 7, pp 229–234. http://www.seaweb.org/resources/articles/writings/safina2.php

WWF (2008) Race for the last bluefin. WWF Mediterranean Project, Zurich, 126 p. https://www.wwf.or.jp/activities/lib/pdf/0811med_tuna_overcapacity.pdf

Zarrad R, Missaoui H (2017) Update of CPUE bluefin tuna, *Thunnus thynnus* (L. 1758) caught by Tunisian purse seiners in the Central Mediterranean. Col Vol Sci Pap ICCAT 73(6):2183–2187

Chapter 2
Some Characteristics of the Bluefin Tuna. Its Geographical Distribution, Areas and Fishing Systems

Abstract Some characteristics of bluefin tuna are described relating to its biology (growth, reproduction, migrations and ethology), physiology, geographical distribution and fishing. Regarding the latter, the evolution of bluefin tuna fisheries over the last seven decades is analyzed, with descriptions of the various fleets working in different regions over this period.

The Atlantic bluefin tuna belongs to the family of the Scombrid fishes (Scombridae) (Collette and Nauen 1985). It can weigh over 725 kg (Crane 1936; Lebedeff 1936; Heldt 1938), reach lengths of 3.3 m (Cort et al. 2013) and live over thirty years (Neilson and Campana 2008). In general, during its fattening phase in the first year of life it reaches 53 cm (4 kg); at age 10, 204 cm (170 kg); at 20 years, 273 cm (410 kg); and at 30 years, 301 cm (550 kg). The official record of the largest ABFT captured in the western Atlantic is 679 kg, a fish caught in Nova Scotia waters (Canada) in 1979 (Fraser 2008). This catch also stands as the current Guinness world record. It forms large shoals and feeds mainly on other fishes, cephalopods, small crustaceans such as *krill* (Euphausiacea) and pelagic crabs, *Polybius henslowii* (Leach), (Estrada et al. 2005; Sarà and Sarà 2007; Logan et al. 2010). Its shape is highly hydrodynamic, as it is entirely adapted to mobility.

ABFT inhabits temperate waters of the North Atlantic and Mediterranean Sea (ICCAT 2010). It is found in the eastern Atlantic from Senegal (Ngon Sow and Ndaw 2010) and Cabo Verde (15° N), the Mediterranean and Black Sea (Zaitsev 2003), almost as far as the Arctic Circle (75° N) where temperatures of 5 °C are recorded (De Metrio et al. 2002; MacKenzie and Myers 2007; Di Natale 2012a), and in the western part from Brazil (Takeuchi et al. 1999, 2009) to Newfoundland (Hurley and Iles 1980). It is also found in the southern Atlantic (Di Natale et al. 2013).

Its bloodstream forms the core of a highly evolved heat exchange system, and so its internal temperature can be maintained at up to 21 °C higher than that of the water surrounding it (Carey et al. 1969; Carey and Lawson 1973). This is one of the reasons for its wide distribution in the ocean. ABFT can appear in the warm waters of the Bahamas at close to 30 °C (Rivas 1954) and 50 days later in Norwegian waters, where the temperature hardly rises above 10 °C (Mather III 1962).

J. L. Cort and P. Abaunza, *The Bluefin Tuna Fishery in the Bay of Biscay*, SpringerBriefs in Biology, https://doi.org/10.1007/978-3-030-11545-6_2

ABFT migrations depend on the age and length of fishes and are mainly related to spawning and the search for food. This behavioural change confirms the theory of the Norwegian scientist J. Hamre (cited by Tiews 1963), which states that ABFT behaviour changes with age, as was eventually shown years later (Lutcavage et al. 2013).

Migrations of adult fishes (fishes > 40 kg) towards spawning areas in the Mediterranean and their return to the ocean for feeding have been known for thousands of years. Today thanks to electronic tagging (Tensek et al. 2018) great advances have been made on this important facet of its biology. Its migrations get longer as its size increases. In order to spawn the tunas emigrate in great shoals (Arena 1979) that choose the most appropriate areas depending on numerous ecological and environmental variables (Alemany et al. 2010).

In general, ABFT migrations, both in adults and juveniles, appear to be associated with large oceanic current systems (Fig. 2.1). Mather III et al. (1995) presented the results of diverse recoveries of large spawning tunas tagged in the Bahamas and recovered a few weeks later in Norwegian waters, which fits this theory as they appear to have followed the Gulf Stream to the north.

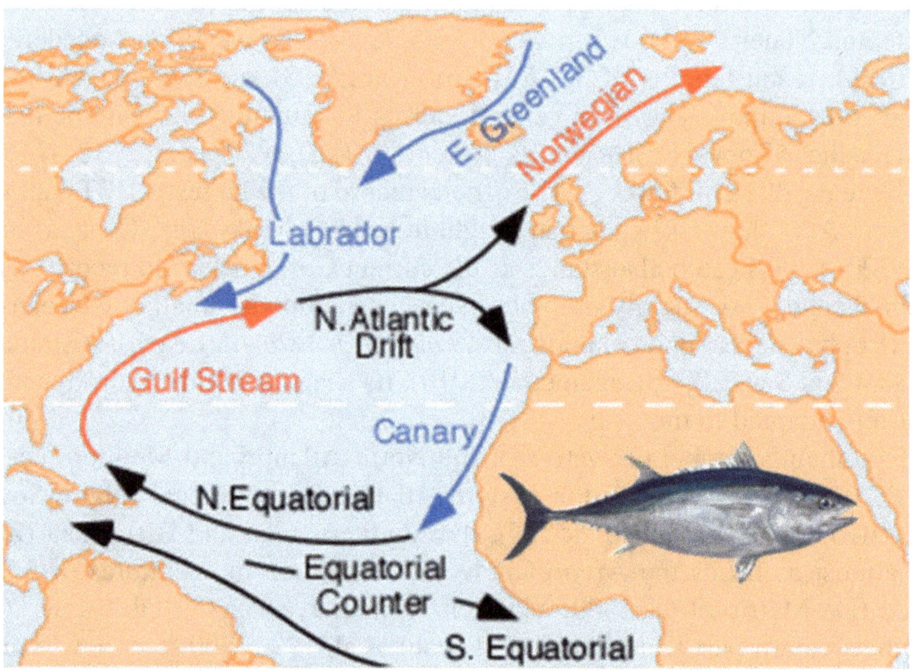

Fig. 2.1 Distribution of marine currents. https://commons.wikimedia.org/wiki/File:Corrientes-oceanicas.png. Red arrows: warm currents; Blue arrows: cold currents; Black arrows: general surface currents

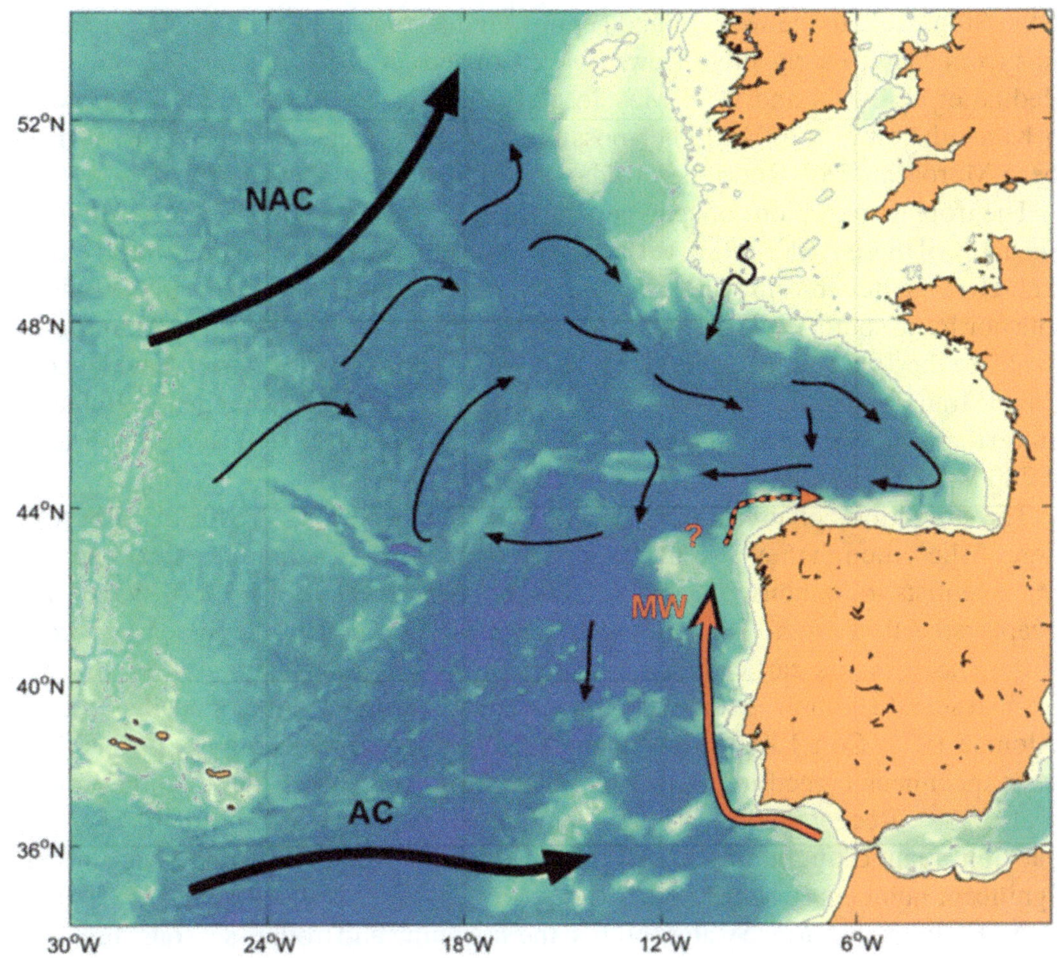

Fig. 2.2 Distribution of currents in the Northeast Atlantic. *NAC* North Atlantic current; *AC* Azores current; *MW* Mediterranean water

There are also juvenile fishes tagged off the eastern coasts of the U.S.A. whose transatlantic migrations from the western to the eastern Atlantic (Mather III et al. 1967) can be related to the North Atlantic Current (NAC), Fig. 2.2; adapted from González-Pola et al. (2005).

Rodríguez-Roda (1967) determined the age at which they reach the first spawn and the age of absolute fecundity. According to this author, ABFT reach full maturity at 5 years of age (130 cm, 50 kg), the age at which a female lays 5 million eggs. At 13 years (230 cm, 250 kg) it lays 30 million eggs. Before him, however, other Portuguese and Italian scientists had established the first sexual maturity at age 3 years and full sexual maturity at age 4 (Frade 1935; Sarà 1961; Scaccini 1965). These data have been confirmed by other authors for the Mediterranean (Piccinetti et al. 2013).

In the West Atlantic there are two spawning areas, one in the Gulf of Mexico (Richards 1977; Mather et al. 1995; Rooker et al. 2007; ICCAT 2010) and another in the Slope Sea (Northwest Atlantic) according to Richardson et al. (2016). Spawning extends from May to early June in the Gulf of Mexico and between June and August in the Slope Sea. In the western and central Mediterranean Sea the spawning period runs from May to early July (Tiews 1963; Corriero et al. 2005; García et al. 2005; Medina et al. 2002) and from May to early June in the Levantine Sea according to Karakulak et al. (2004). Other possible spawning areas have been cited in the Ibero-Moroccan Bay (Rodríguez-Roda 1969; Di Natale et al. 2017).

The trophic migrations of spawners begin once the spawning period has finished and many of these ABFT return to the Atlantic Ocean (Rodríguez-Roda 1964; Tensek et al. 2018). After crossing the Strait of Gibraltar, the shoals disperse and fishes head both north and south between June and December (De Metrio et al. 2002; Aranda et al. 2013). In the 1960s North American scientists (Mather III 1962) found evidence of 'direct' transatlantic migrations of large spawning bluefin tunas tagged in Bahamas (June) that were recovered in Norwegian waters 50 days after, indicating a migration of 7800 km (155 km/day).

The migrations of juvenile fishes (<40 kg) are generally shorter than those of larger fishes, though they also make transatlantic migrations in both directions. The first scientists to demonstrate this were North Americans in the 1960s, when they recaptured ABFT in the Bay of Biscay that had been tagged off the eastern coast of the U.S.A. years earlier (Mather III et al. 1967). Soon afterwards French and Spanish scientists showed that this migration also occurred in the opposite direction (Aloncle 1973; Cort 1990). In recent years, in projects financed by ICCAT-GBYP, other specimens tagged in the Bay of Biscay migrated across the Atlantic (Tensek et al. 2018). Recent studies on the chemical composition of otoliths have shown that transatlantic migrations of ABFT juveniles take place in some years in very significant numbers (Rooker et al. 2014).

ABFT frequent surface waters both in the spawning and trophic seasons (Fig. 2.3). Electronic tagging studies also reveal that they often dive to great depths, sometimes to over 800 m (Block et al. 2005).

The evolution of bluefin tuna fishing in the north Atlantic has gone through different phases over the last seven decades (Fig. 2.4). In this figure it can be seen how fishing for this species has changed since 1950, the first year of the ICCAT data base. It can be seen how in the 1950s they were mainly caught in the eastern Atlantic with TR (Cort et al. 2012; Pereira 2012), PS (Hamre 1960; Tangen 2009) and BB (Le Gall 1951; Cort 1990), and in the Mediterranean with TR (Di Natale 2012b) and PS (Piccinetti 1980); in the western Atlantic there were catches with TR (Dean et al. 2012), recreational, small nets and harpoon (Hurley and Iles 1980). The most remarkable in the 1960s was the development of the Japanese LL, which stretched as far as the southern hemisphere off Brazil (Takeuchi et al. 2009) and also PS in western fisheries (Sakagawa 1975; Porch 2005). In the 1970s fishing using TR diminished in the Strait of Gibraltar (Rodríguez-Roda 1977) and in the Mediterranean (ICCAT 2010), whereas the use of PS fell in the north of Europe (Nøttestad et al. 2009). In the same decade PS (Farrugio 1980) and LL (Rey 1980) developed in the Mediterranean.

Fig. 2.3 Bluefin tuna in trophic migration *Artist*: Lineke Zubieta (Santander, Spain) (Documentary archive, *IEO*)

In the 1980s PS disappeared in the north of Europe (Nøttestad et al. 2009) while it continued to proliferate in the Mediterranean (Farrugio 1980; Arena 1982, 1988). From the 1990s the use of PS continued to increase in the western Mediterranean (Liorzou and Bigot 1993; De la Serna and Platonenko 1996) and eastern Mediterranean (Libya and Turkey) according to ICCAT (2010) and Karakulak (2003). Japanese longline also increased in the central and eastern Atlantic (Kimoto and Itoh 2017). These modern systems took over in the last century to become the most commonly used gears (ICCAT 1980). In more recent years (2000s) the greatest fishing effort is exerted in the Mediterranean (Fig. 2.5), where purse seine is used to catch fishes for fattening farms (Deguara et al. 2010; Galaz 2012; Hattour and Kouched 2013; Gordoa 2014, 2017), though in the Strait of Gibraltar fishing with traps continues (De la Serna et al. 2012; Santos et al. 2016).

Figure 2.6 shows the panorama of ABFT fishing in the middle of the 20th century when the fisheries, which had renewed their activities following the Second World War, were at their height, as was the case of the northern European fisheries and a little

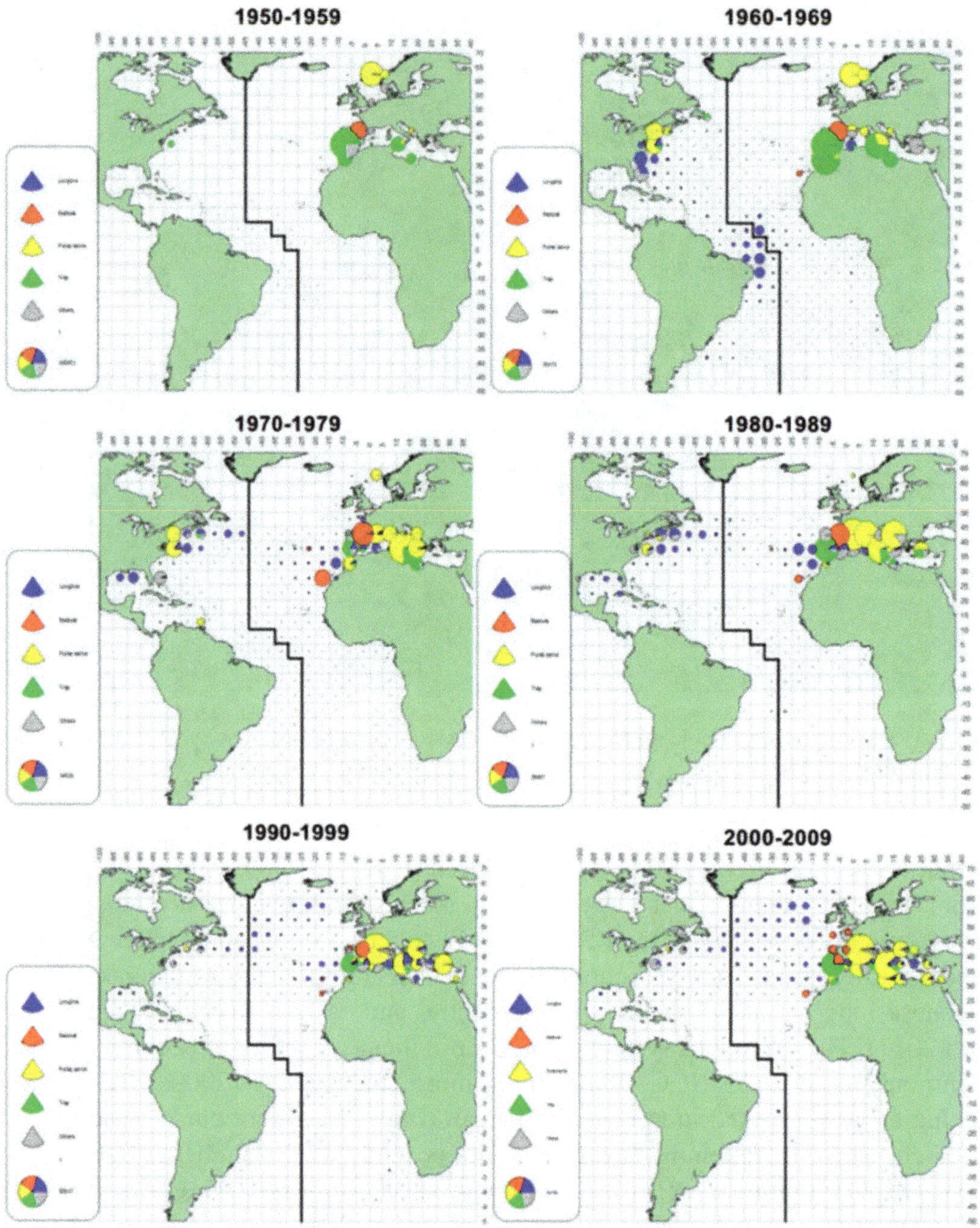

Fig. 2.4 Distribution of bluefin tuna fishing by decade [Blue, longline (LL); Red, baitboat (BB); Yellow, purse seine (PS); Green, trap (TR); Grey, others] (Courtesy of ICCAT)

Fig. 2.5 Modern Spanish purse seiner in the western Mediterranean (Courtesy of *Grup Balfegó*)

later the Asian longline fisheries, mainly Japan. Also included are the purse seine fleets of the Mediterranean, whose rapid development began in the 1960s (Piccinetti 1980). The figure provides a vision of ABFT fishing when it was at its height and also its subsequent decline, using red circles to represent the fisheries that abandoned their activities following the overfishing of the 1960s in the eastern part of the Atlantic and Mediterranean, and in the 1970s in the western part of the Atlantic Ocean.

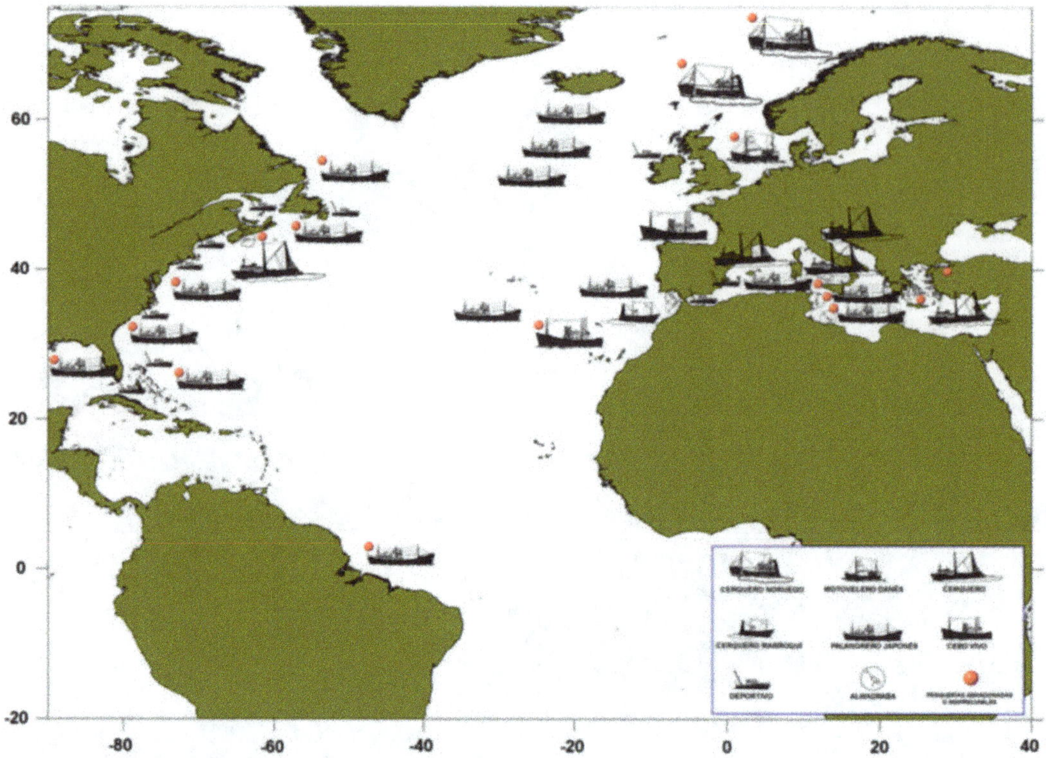

Fig. 2.6 Fleets both present and past targeting bluefin tuna. Small artisanal fisheries are not included (explanation in the text). Legend: Cerquero noruego = Norwegian purse seiner; Motovelero danés = Danish motorsailer; Cerquero = Purse seiner; Cerquero marroquí = Moroccan purse seiner; Palangrero japonés = Japanese long liner; Cebo vivo = Bait boat; Deportivo = Sportive; Almadraba = Tuna trap; Red bullet = Abandoned or almost unworkable fisheries

References

Alemany F, Quintanilla L, Vélez-Belchí P, García A, Cortés D, Rodríguez JM et al (2010) Characterization of the spawning habitat of Atlantic bluefin tuna and related species Balearic Sea (western Mediterranean) Prog Ocean 86:21–38

Aloncle H (1973) Marquage de thons dans le Golfe de Gascogne. Col Vol Sci Pap ICCAT 1:445–458

Aranda G, Abascal FJ, Varela JL, Medina A (2013) Spawning behaviour and post-spawning migration patterns of Atlantic Bluefin tuna (Thunnus thynnus) ascertained from satellite archival tags. PLoS ONE 8(10):e76445. https://doi.org/10.1371/journal.pone.0076445

Arena P (1979) Aspects biologiques et comportement des concentrations génétiques du thon rouge en Mediterranée. Actes Colloq CNEXO 8:53–57

Arena P (1982) La pêche à la senne tournante du thon rouge (*Thunnus thynnus* L.) dans les bassins maritimes occidentaux italiens. Col Vol Sci Pap ICCAT 17(2):281–292

Arena P (1988) Rilevazini e studi sulle affluenze del tonno nel Tirreno e sull´andamento della pesca da parte delle "tonnare volanti" nel quadriennio 1984–1988. E.S.P.I. Ente Siciliano per la Promozione Industriale-Palermo Unità Operativa n. 5, 66 p

Block BA, Teo SLH, Walli A, Boustany A, Stikesbury MJW, Farwell CJ, Weng KC, Dewar H, Williams TD (2005) Electronic tagging and population structure of Atlantic bluefin tuna. Nature 434:1121–1127

Carey F, Teal JM (1969) Regulation of body temperature by the bluefn tuna. Comp Biochem Physiol 28:205–213

Carey FG, Lawson KD (1973) Temperature regulation in free-swimming bluefin tuna. Comp Biochem Physiol 44:275–292

Collette BB, Nauen CE (1985) FAO species catalogue. Vol. 2, Scombrids of the world. An annotated and illustrated catalogue of Tunas, Mackerels, Bonitos and Related Species Known to Date. FAO Fisheries Synopsis No 125. Vol. 2-Mit 81 figs., 137 pp. Rome: FAO 1983. ISBN-Nr. 92-5-101381-0

Corriero A, Kakakulak S, Santamaría N, Deflorio M, Spedicato D, Addis P, Fenech-Farrugia A, Vassallo-Agius R, de la Serna JM, Oray I, Cau A, De Metrio G (2005) Size and age at sexual maturity of female bluefin tuna (*Thunnus thynnus* L. 1758) from the Mediterranean Sea. J Appl Ichthyol 21:483–486

Cort JL (1990) Biología y pesca del atún rojo, *Thunnus thynnus* (L.), del mar Cantábrico. Publicaciones especiales, IEO 4:272 p

Cort JL, de la Serna JM, Velasco M (2012) El peso medio del atún rojo (*Thunnus thynnus*) capturado por las almadrabas del sur de España entre 1914–2010. Col Vol Sci Pap ICCAT 67:231–241

Cort JL, Deguara S, Galaz T, Mèlich B, Artetxe I, Arregi I et al (2013) Determination of $L_{\max}$ for Atlantic Bluefin Tuna, *Thunnus thynnus* (L.), from meta-analysis of published and available biometric data. Rev Fish Sci 21(2):181–212 https://doi.org/10.1080/10641262.2013.793284

Crane J (1936) Notes on the biology and ecology of giant tunas, *Thunnus thynnus* Linnaeus, observed at Portland. Maine Zool 21(16):207–212

Dean JM, Andrushchenko I, Neilson JD (2012) The western Atlantic bluefin tuna trap fishery. Col Vol Sci Pap ICCAT 67:309–321

Deguara S, Caruana S, Agius C (2010) An appraisal of the use of length–weight relationships to determine growth in fattened Atlantic bluefin tuna, *Thunnus thynnus*. Col Vol Sci Pap ICCAT 65:776–781

De la Serna JM, Platonenko S, Alot E (1996) Observaciones preliminares sobre las capturas de atún rojo (*Thunnus thynnus*) con artes de cerco en el Mediterráneo occidental. Col Vol Sci Pap ICCAT 45(2):117–122

De la Serna JM, Macías D, Ortiz de Urbina JM, Rodríguez-Marín E, Abascal F (2012) Study on eastern Atlantic and Mediterranean Bluefin tuna stocks using Spanish traps as scientific observatories. Col Vol Sci Pap ICCAT 67:331–343

De Metrio G, Arnold GP, Block BA, de la Serna JM, Deflorio M, Cataldo M, Yannopoulos C, Megalofonou P, Beeper S, Farwell C, Seitz A (2002) Behaviour of post-spawning Atlantic Bluefin tuna tagged with pop-up satellite tags in the Mediterranean and eastern Atlantic. Col Vol Sci Pap ICCAT 54(2):415–424

Di Natale A (2012a) New data on the historical distribution of bluefin tuna (*Thunnus thynnus*, L.) in the Artic Ocean. Col Vol Sci Pap ICCAT 68(1):102–114

Di Natale A (2012b) Literature on eastern Atlantic and Mediterranean tuna trap fishery. Col Vol Sci Pap ICCAT 67:175–220

Di Natale A, Idrissi M, Justel Rubio A (2013) The mistery of bluefin tuna (*Thunnus thynnus*) presence and behavior in Central-South Atlantic in recent years. Col Vol Sci Pap ICCAT 69(2):857–868

Di Natale A, Tensek S, Pagá García A (2017) ICCAT Atlantic-wide research programme for bluefin tuna (GBYP): activity report for the last part of phase 5 and first part of phase 6 (2015–2016). Col Vol Sci Pap ICCAT 73(7):2424–2503

Estrada JA, Lutcavage M, Thorrold SR (2005) Diet and trophic position of Atlantic bluefin tuna (*Thunnus thynnus*) inferred from stable carbon and nitrogen isotope analysis. Mar Biol 147:37–45. https://doi.org/10.1007/s00227-004-1541-1

Farrugio H (1980) Bluefin fishing in the French area of the Mediterranean. Development and characteristics. Col Vol Sci Pap ICCAT 11:98–117

Frade F (1935) Recherches sur la maturité sexuelle du Thon rouge. C R Congr Int Zool 12

Fraser K (2008) Possessed: World record holder for bluefin tuna. Kingstown, Nova Scotia: T & S Office Essentials and printing, 243 p

Galaz T (2012) Eleven years—1995–2005—of experience on growth of bluefin tuna (*Thunnus thynnus*) in farms. Col Vol Sci Pap ICCAT 68(1):163–175

García A, Alemany F, de la Serna JM, Oray I, Karakulak S, Rollandi L, Arigo A, Mazzola S (2005) Preliminary results of the 2004 bluefin tuna larval surveys off different Mediterranean sites (Balearic Archipelago, Levantine Sea, and the Sicilian Channel). Col Vol Sci Pap ICCAT 58:1420–1428

González-Pola C, Lavín A, Vargas-Yáñez M (2005) Intense warming and salinity modification of intermediate water masses in the southeastern corner of the Bay of Biscay for the period 1992–2003. J Geophys Res 110:C05020. https://doi.org/10.1029/2004jc002367

Gordoa A (2014) Catch rates and catch size structure of the Balfegó purse seine fleet in Balearic waters from 2000 to 2014; two years of size frequency distribution based on video techniques. Col Vol Sci Pap ICCAT 71(4):1803–1812

Gordoa A (2017) Updated bluefin tuna CPUE and catch structure from Balfegó purse seine fleet in Balearic waters from 2000 to 2016. Col Vol Sci Pap ICCAT 73(6):2020–2025

Hamre J (1960) Tuna investigation in Norwegian coastal waters 1954-1958. Ann Biol Cons Int Expl Mer 15:197–211

Hattour A, Kouched W (2013) (Temporal distribution of size and weight of fattened bluefin tuna (*Thunnus thynnus* L.) from Tunisian farms (2005–2010). *Mediterranean Marine Science*. (2013). http://www.medit-mar-sc.net, http://dx.doi.org/10.12681/mm.513

Heldt H (1938) Le thon rouge et sa pêche 10 Rapp. Comm Internat Explor Medit 11:311–358

Hurley PCF, Iles TD (1980) A brief description of Canadian fisheries for Atlantic bluefin tuna. Col Vol Sci000 Pap. ICCAT 11:93–97

ICCAT (1980) Report of bluefin tuna workshop (Santander, Spain, 3–8 Sept 1979). Col Vol Sci Pap ICCAT 11:1–89. https://www.iccat.int/Documents/CVSP/CV011_1980/CV011000001.pdf

ICCAT (2010) ICCAT manual. Description of species. Chapter 2; 2.1.5 Atlantic bluefin tuna, vol 99. Madrid, ICCAT, pp 93–111. http://iccat.int/Documents/SCRS/Manual/CH2/2_1_5_BFT_ENG.pdf

Karakulak S (2003) Bluefin tuna fishery in Turkey. In: Workshop on farming, management and conservation of bluefin tuna, 5–7 April 2003. Istanbul, Turkey, Turkish Marine Research Foundation Publication No. 13, pp 120–133

Karakulak S, Oray I, Corriero A, Deflorio M, Santamaria N, Desantis S, De Metrio G (2004) Evidence of a spawning area for the bluefin tuna (*Thunnus thynnus*) in the Eastern Mediterranean. J Appl Ichthyol 20:318–320

Kimoto A, Itoh T (2017) Simple update of the standardized bluefin tuna CPUE of Japanese longline fishery in the Atlantic up to 2016 fishing year. Col Vol Sci Pap 73(6):1957–1976

Lebedeff W (1936) Paradise for big game fishing. Tunny, 700 kgs; Swordfish, 180 kgs; Shark, 1800 kgs. Winter season 1935–36 in Turkey. Fish Gaz 113(3102):420–421

Le Gall J (1951) Le thon rouge (*Thunnus thynnus* L.) dans le Golfe de Gascogne en 1951. Ann Biol Cons Int Expl Mer 8:82–83

Liorzou B, Bigot JL (1993) Actualisation des données sur le thon rouge exploité au large des cotes françaises de Méditerranée. Col Vol Sci Pap ICCAT 40(1):302–306

Logan JM, Rodríguez-Marín E, Goñi N, Barreiro S, Arrizabalaga H, Golet WJ, Lutcavage M (2010) Diet of young Atlantic bluefin tuna (*Thunnus thynnus*) in eastern and western foraging gound. Mar Biol https://doi.org/10.1007/s00227-010-1543-0, 12 p

Lutcavage M, Galuardi B, Lam TCH (2013) Predicting potential Atlantic spawning grounds of western Atlantic bluefin tuna based on electronic tagging results, 2002–2011. Col. Vol. Sci. Pap. ICCAT 69:955–961

MacKenzie BR, Myers RA (2007) The development of the northern European fishery for north Atlantic bluefin tuna (*Thunnus thynnus*) during 1900–1950. Fish Res https://doi.org/10.1016/j.fishres.2007.01.013

Mather III FJ. Distribution and migrations of North Atlantic bluefin tuna. In: Proccedings of the seveth international game fish conference, Galveston, T exas, 15 Nov (1962)

Mather III FJ, Bartlett MR, Beckett JS (1967) Transatlantic migrations of young bluefin tuna. J Fish Res Bd Canada 24(9):1991–1997

Mather III FJ, Mason Jr JM, Jones AC (1995) Historical document: life history and fisheries of Atlantic bluefin tuna. NOAA technical memorandum, NMFS-SEFSC-370, Miami Fl, 165 p

Medina A, Abascal FJ, Megina C, García A (2002) Stereological assessment of the reproductive status of female Atlantic northern bluefin tuna during migrations to Mediterranean spawning grounds through the Strait of Gibraltar. J Fish Biol 60:203–217. https://doi.org/10.1111/j.1095-8649.2002.tb02398.x

Neilson J, Campana SE (2008) A validated description of age and growth of Western Atlantic bluefin tuna (*Thunnus thynnus*). Can J Fish Aquat Sci 65:1523–1527

Ngom Sow F, Ndaw S (2010) Bluefin tuna caught by Spanish baitboat and landed in Dakar in 2010. Col Vol Sci Pap ICCAT 66(2):883–887

Nøttestad L, Tangen Ø, Sundby S (2009) Norwegian fisheries since the early 1960s: what went wrong and what can we do? Col Vol Sci Pap ICCAT 63:231–232

Pereira J (2012) Historical bluefin tuna catches from southern Portugal traps. Col Vol Sci Pap ICCAT 67:88–105

Piccinetti C (1980) The bluefin seine fishery in the Adriatic. Col Vol Sci Pap ICCAT 11:124–128

Piccinetti C, Di Natale A, Arena P (2013) Eastern bluefin tuna (*Thunnus thynnus*, L.) reproduction and reproductive areas and season. Col Vol Sci Pap ICCAT 69(2):891–912)

Porch CE (2005) The sustainability of western Atlantic bluefin tuna: a warm blooded fish in a hot blooded fishery. Bull Mar Sci 76:363–384

Rey JC (1980) Description of the longline fishery in the Spanish Mediterranean. Col Vol Sci Pap ICCAT 11:178–179

Richards WJ (1977) A further note on Atlantic bluefin tuna spawning. Col Vol Sci Pap ICCAT 6(2):335–336

Richardson DE, Marancik KE, Guyon JR, Lutcavage ME, Galuardi B, Lam CH, Walsh HJ, Wildes S, Yates DA, Hare JA (2016) Discovery of spawning ground reveals diverse migration strategies in Atlantic bluefin tuna (*Thunnus thynnus*). PNAS 113(12):3299–3304

Rivas LR (1954) A preliminary report on the spawning of the western North Atlantic bluefin tuna (*Thunnus thynnus*) in the straits of Florida. Bull Mar Sci Gulf Caribb 4(4):302–322

Rodríguez-Roda J (1964) Biología del atún, *Thunnus thynnus* (L.), de la costa sudatlántica española. Inv Pesq 25:33–146

Rodríguez-Roda J (1967) Fecundidad de atún, *Thunnus thynnus* (L.), de la costa sudatlántica española. Inv Pesq 31(1): 33–52

Rodríguez-Roda J (1969) Los atunes jóvenes y el problema de sus capturas masivas. Publicaciones Técnicas de la Junta de Estudios de Pesca. Subsecretaría de la Marina Mercante 8:159–162

Rodríguez-Roda J (1977) Análisis de la población de atunes, *Thunnus thynnus* (L.), capturados por la almadraba de Barbate (Golfo de Cádiz) durante los años 1963 a 1975. Inv Pesq 41(2):263–273

Rooker J, Alvarado J, Block B, Dewar H, De Metrio G, Prince E, Rodríguez-Marín E, Secor D (2007) Life and stock structure of Atlantic bluefin tuna (*Thunnus thynnus*). Rev Fish Sci 15:265–310

Rooker J, Arrizabalaga H, Fraile I, Secor DH, Dettman DL, Abid N, Addis P, Deguara S, Karakulak FS, Kimoto A, Sakai O, Macias D, Santos MN (2014) Crossing the line: migratory and homing behaviours of Atlantic bluefin tuna. Mar Ecol Prog Ser 504:265–276

Sakagawa G (1975) The purse-seine fishery for bluefin tuna in the Northwestern Atlantic Ocean. Mar Fish Rev 1126 37(3):8

Santos MN, Rosa D, Coelho R, Lino PR (2016) New observations on the bluefin tuna trap fishery off southern Portugal (NE Atlantic) between 1998–2014: trends on potential catches, catch-at-size and sex ratios. Col Vol Sci Pap ICCAT 72(5):1350–1364

Sarà R (1961) Observations systématiques sur la presence du thon dans les madragues siciliennes. Proc Gen Fish Coun Médit 6:41–61

Sarà G, Sarà R (2007) Feeding habits and trophic levels of bluefin tuna *Thunnus thynnus* of different size classes in the Mediterranean Sea. J Appl Ichthyol 23(2):122–127

Scaccini A (1965) Biologia e pesca dei tonni nei mari italiani. Ministero Marina Mercantile, Mem 12:101

Takeuchi Y, Suda A, Suzuki Z (1999) Review of information on large bluefin tuna caught by Japanese longline fishery off Brasil from the late 1950s to the early 1960s. Col Vol Sci Pap ICCAT 49(2):416–428

Takeuchi Y, Oshima K, Suzuki Z (2009) Inference on the nature of Atlantic bluefin tuna off Brazil caught by the Japanese longline fishery around the early 1960s. Col Vol Sci Pap ICCAT 63:186–194

Tangen M (2009) The Norwegian fishery for Atlantic bluefin tuna. Col Vol Sci Pap ICCAT 63:79–93

Tensek S, Pagá García A, Di Natale A (2018) ICCAT GBYP tagging activities in phase 6. Col Vol Sci Pap ICCAT 74(6):2861–2872

Tiews K (1963) Synopsis of biological data on bluefin tuna, *Thunnus thynnus* (Atlantic and Mediterranean). FAO Fish Rep 6(2):422–481

Zaitsev Y (2003) Bluefin tuna in the Black Sea. In: Workshop on farming, management and conservation of bluefin tuna, 5–7 April 2003. Istanbul, Turkey, Turkish Marine Research Foundation Publication No. 13, pp 118–119

Chapter 3
Two Stocks

Abstract The management of bluefin tuna resources within ICCAT is based on the consideration of two distinct stocks. A definition is provided of what a stock is from the point of view of a species subjected to fishing exploitation and the basic criterion by which the separation of bluefin tuna into western and eastern stocks was adopted.

For the purposes of resource management, the International Commission for the Conservation of Atlantic Tunas (ICCAT) considers the population of the North Atlantic as two stocks[1]: the eastern, which includes the Mediterranean, and the western (ICCAT 2010). These are managed separately. The line defining the separation of the stocks is the 45° W meridian (Fig. 3.1). The basis on which this separation of eastern and western stocks is made is the fact that there are two main spawning areas, one in the Gulf of Mexico and the other in the Mediterranean.

This dividing line in no way means that ABFT cannot cross it; they do so without any difficulty as demonstrated in several articles: Block et al. (2005), Boustany et al. (2008), Fraile et al. (2014), Rooker et al. (2014), Puncher (2015), Arregui et al. (2018), and Tensek et al. (2018). Mixing of the two stocks of ABFT is variable in quantity and in time (Siskey et al. 2016). In spite of this, ABFT in most cases shows fidelity to one spawning ground by returning to its birthplace (Block et al. 2005).

Five synopses on this species have been published in which other aspects are detailed that have not been considered in the present article: Tiews (1963), Mather et al. (1973, 1995), Fromentin and Powers (2005), Rooker et al. (2007) and ICCAT (2010).

[1] Stock: A stock constitutes a biological unit of a species that forms a group with similar ecological characteristics, the unit being what is subject to assessment and management.

J. L. Cort and P. Abaunza, *The Bluefin Tuna Fishery in the Bay of Biscay*, SpringerBriefs in Biology, https://doi.org/10.1007/978-3-030-11545-6_3

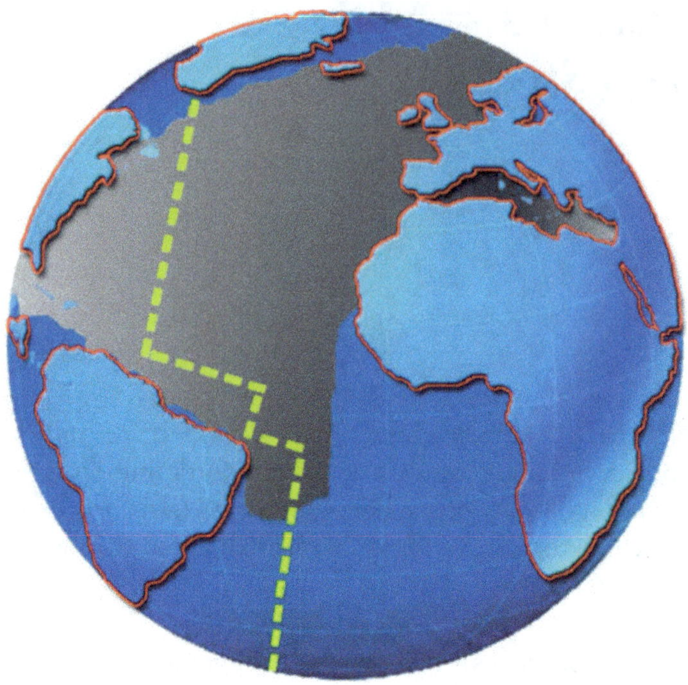

Fig. 3.1 Dividing line of the bluefin tuna stocks (45º West). (Taken from Cort et al. 2010)

References

Arregui I, Galuardi B, Goñi N, Lam CH, Fraile I, Santiago J, Lutcavage M, Arrizabalaga H (2018) Movements and geographic distribution of juvenile bluefin tuna in the Northeast Atlantic, described through internal and satellite archival tags. ICES J Mar Sci. https://doi.org/10.1093/icesjms/fsy056

Block BA, Teo SLH, Walli A, Boustany A, Stikesbury MJW, Farwell CJ, Weng KC, Dewar H, Williams TD (2005) Electronic tagging and population structure of Atlantic bluefin tuna. Nature 434:1121–1127

Boustany AM, Reeb CA, Block BA (2008) Mitochondrial DNA and electronic tracking reveal population structure of Atlantic bluefin tuna (*Thunnus thynnus*). Mar Biol 156:13–24. https://doi.org/10.1007/s00227-008-1058-0

Cort JL, Abascal F, Belda E, Bello G, Deflorio M, de la Serna JM, Estruch V, Godoy D, Velasco M (2010) ABFT tagging manual of the Atlantic-wide research programme for bluefin tuna. ICCAT, 47 p. https://www.iccat.int/GBYP/Docs/Tagging_Manual.pdf

Fraile I, Arrizabalaga H, Rooker R (2014) Origin of Atlantic bluefin tuna (*Thunnus thynnus*) in the Bay of Biscay. ICES J Mar Sci 72(2):625–634. https://doi.org/10.1093/icesjms/fsu156

Fromentin JM, Powers J (2005) Atlantic bluefin tuna: population dynamics, ecology, fisheries and management. Fish and Fisheries 6:281–306

ICCAT (2010) ICCAT Manual. Description of species. Chapter 2; 2.1.5 Atlantic Bluefin Tuna, vol 99, pp 93–111. ICCAT, Madrid. http://iccat.int/Documents/SCRS/Manual/CH2/2_1_5_BFT_ENG.pdf

Mather III FJ, Mason Jr JM, Jones AC (1973) Distribution fisheries and life history data relevant to identification of Atlantic bluefin tuna stocks. ICCAT Col Vol Sci Pap 2:234–258

Mather III FJ, Mason Jr JM, Jones AC (1995) Historical document: life history and fisheries of Atlantic bluefin tuna. NOAA Technical Memorandum, NMFS-SEFSC-370, Miami Fl, 165 pp.

Puncher GN (2015) Assessment of the population structure and temporal changes in spatial dynamics and genetic characteristics of Atlantic bluefin tuna under a fishery independent framework. Doctoral thesis submitted to Alma Mater Studiorum - Università di Bologna & Universiteit Gent, 237 pp. http://www.iccat.int/GBYP/Documents/BIOLOGICAL%20STUDIES/Scientific_Papers/Puncher_PhD_Thesis.pdf

Rooker J, Alvarado J, Block B, Dewar H, De Metrio G, Prince E, Rodríguez-Marín E, Secor D (2007) Life and stock structure of Atlantic Bluefin Tuna (*Thunnus thynnus*). Rev Fish Sci 15:265–310

Rooker J, Arrizabalaga H, Fraile I, Secor DH, Dettman DL, Abid N, Addis P, Deguara S, Karakulak FS, Kimoto A, Sakai O, Macias D, Santos MN (2014) Crossing the line: migratory and homing behaviours of Atlantic bluefin tuna. Mar Ecol Prog Ser 504:265–276

Siskey MR, Wilberg MJ, Allman RJ, Bernett BK, Secor DH (2016) Forty years of fishing: changes in age structure and stock mixing in northwestern Atlantic bluefin tuna (*Thunnus thynnus*) associated with size-selective and long-term exploitation. ICES J Mar Sci. https://doi.org/10.1093/icesjms/fsw115

Tensek S, Pagá García A, Di Natale A (2018) ICCAT GBYP tagging activities in phase 6. ICCAT Col. Vol. Sci. Pap. 74(6):2861–2872

Tiews K (1963) Synopsis of biological data on bluefin tuna, *Thunnus thynnus* (Atlantic and Mediterranean). FAO Fish Rep 6(2):422–481

Chapter 4
The Bluefin Tuna Catch in the Strait of Gibraltar. A Review of Its History

Abstract Taking bluefin tuna in the Strait of Gibraltar as a reference, a description is made of the fishing methods used in its capture from the ancient populations of hominids to the present day. To do so, and based on recent paleoanthropological studies, the hypothetical way in which this fish would have been caught by the neanderthals is described (over 30,000 years ago); based on an extensive bibliography on the subject, the way fishing would have taken place in the Roman city of *Baelo Claudia* (200 B.C.) is described; and how it was during the Modern Age and is nowadays. Information is given on the fishing statistics from three Spanish traps between 1525 and 1756, upon which different scientists have pronounced in recent publications, and emphasis is placed on the overfishing that has taken place since the middle of the 20th century, its consequences, and how these have been overcome.

4.1 Over 30,000 Years Ago

The caves of the Rock of Gibraltar offer much evidence of the presence of hominid populations, which have been studied since the mid-nineteenth century. An international group of scientists intensified digs from 1995 in the Gorham (Fig. 4.1) and Vanguard caves (Giles-Pacheco et al. 2001). The following phrase is taken from these studies: "*En los últimos registros paleontológicos hemos detectado presencia de macro-ictiofauna identificable por vértebras de túnidos de medio y gran tamaño*" (*In the last paleontological records we have detected the presence of macro-ichthyofauna identifiable by tuna vertebrae of medium and large size*). This undoubtedly refers to ABFT samples found in stratum III corresponding to the Magdalenian period (15,000 years ago) in the upper Paleolithic; although these caves are known to have been inhabited more than 40,000 years ago (Middle Paleolithic).

In later excavations (1999–2005; Fig. 4.2) in the Gorham cave (Finlayson et al. 2006) nodules of carbon have been found throughout stratum IV (20,000–30,000 years ago), which is associated with occupation by neanderthals, which in turn means that it was used by these hominids to make fires.

Stringer et al. (2008) discovered that the neanderthals that lived in these caves more than 30,000 years ago were already eating fish. Although consumption of ABFT

J. L. Cort and P. Abaunza, *The Bluefin Tuna Fishery in the Bay of Biscay*, SpringerBriefs in Biology, https://doi.org/10.1007/978-3-030-11545-6_4

Fig. 4.1 Entrance to the Gorham cave (Gibraltar) *Artist*: Lineke Zubieta (Santander, Spain) (Documentary archive, *IEO*)

Fig. 4.2 Scientists excavating inside the Gorham cave (Courtesy of Francisco Giles Pacheco)

by neanderthals is unproven, Cort (2006) cites the sporadic beaching of this species as a result of the stampedes produced in chases by killer whales, *Orcinus orca* (L.), as one of the possible ways in which hominids may have begun to eat it in addition to the small species found in the cave (*Diplodus sargus* (L.) or *D. vulgaris* (F.)). Such events are common on the beaches of the Strait of Gibraltar. The narrowness of the Strait of Gibraltar at that time (10 km), the lower sea level, 120 m below that of the present day (Allen et al. 1999; Tzedakis et al. 2002; Rodríguez-Vidal et al. 2013), with broad valleys (now underwater) and many islands (Montero and Bernal 2011), conditions may well have been favourable to beaching of tunas and other species.

The Strait of Gibraltar 30,000 years ago: from a cave on the edge of the broad plain the members of several neanderthal families descend on their way to the beach which, due to the present glaciation, is several kilometres away as a result of the 120-m fall in the sea level. It is spring, and as they know well from their predecessors, at this time of year the tunas, full of eggs, migrate eastwards. The strait now consists of two distinct broad channels and Africa is only 10 km away. The current is very strong, but the killer whales will soon begin the chase of the shoals of huge tunas. In the chase some of the enormous fishes will be beached, and each of them will ensure a lot of food for these last neanderthal families, who came here two thousand years earlier when they were fleeing from the glacial cold spreading through the European continent. Figures 4.3, 4.4 *and* 4.5 *show three scenes that may have taken place in those times in the Strait of Gibraltar.*

Fig. 4.3 Beached bluefin tuna being collected by the neanderthals in the Strait of Gibraltar 30,000 years ago An interpretation by Cort (2007) *Artist*: Lineke Zubieta (Santander, Spain) (Documentary archive, *IEO*)

Fig. 4.4 Gutted bluefin tuna being transported to the cave An interpretation by Cort (2007) *Artist*: Lineke Zubieta (Santander, Spain) (Documentary archive, *IEO*)

Fig. 4.5 Neanderthals feeding on bluefin tuna An interpretation by Cort (2007) *Artist*: Lineke Zubieta (Santander, Spain) (Documentary archive, *IEO*)

4.2 Phoenician-Roman Empire

Many samples remain from the Phoenician and Roman empires (1200 BC–470 AD) both in the literature and the iconography of the interest and commerce arising from ABFT among coastal populations. Figure 4.6 shows the silhouettes of ABFT found in different regions (Sarà 1998).

For thousands of years bluefin tuna has been caught in the vicinity of the Strait of Gibraltar and throughout the Mediterranean basin (García Vargas and Florido del Corral 2011), but it was the Phoenicians who pioneered the use of nets which, over time, led to the traps, a fishing gear made up of labyrinths that was installed near the coast (Sarà 1998). Di Natale (2012) says: "*trap fishery is the most ancient industrial activity in the fishery sector, because it is well known that "tuna traps" were operating in ancient times in Greece, and tuna traps were very active during Phoenician and Roman times, in many coastal areas*".

During the Roman Empire ABFT formed the basis of considerable commerce both inside and outside the Mediterranean basin. The Roman city of *Baelo Claudia* was in the area of the Strait of Gibraltar (Arévalo and Bernal 2007; Fernández et al. 2007), situated in the inlet of Bolonia within what is now the Natural Park of the Strait just 12 km from the city of Tarifa in the province of Cádiz. The city was founded around the 3rd century BC and reached its height under the emperor Claudio (1st century BC), Fig. 4.7.

Fishing, the salting industry and production of *garum* (a sauce made from the leftovers of ABFT and other fishes macerated in the sun in brine) were the main sources of wealth in *Baelo Claudia*, a city from which exports were sent to the rest of the empire. García Vargas and Bernal (2009) provided plenty of information on the production of salted meats, fish sauces and commerce in the south of *Hispania*,

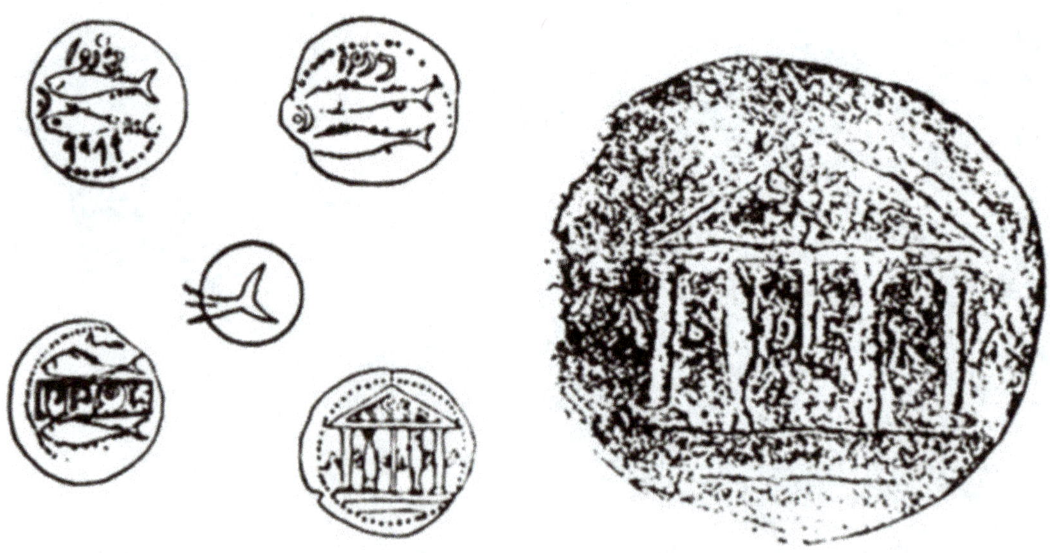

Fig. 4.6 Bluefin tuna on ancient coins. (From Ponsich, 1990. Taken from Sarà, 1998)

Fig. 4.7 Ruins of the Roman city of *Baelo Claudia* (Cádiz, Spain) (Courtesy of Kurt M. Schaefer)

which reflects the importance of ABFT in this industry. Morales and Roselló (2007) published the find of 683 ABFT vertebrae in a refuse tip of the *Baelo Claudia* factory in *Punta del Caramiñal*; and Niveau de Villedary (2009, 2011) even spoke of the meaning that ABFT might have had in the funeral ceremonies of those times, since vertebrae of this fish have been found in a grave pit with human remains (3rd century BC). Cort (2006, 2007), and García Vargas and Florido del Corral (2011) maintain that ABFT fishing in the Strait of Gibraltar in Roman times was conducted using nets thrown from boats. Other references to fishing for ABFT in the ancient world in the vicinity of the Strait of Gibraltar (Algarve, Portugal; Lixus, Morocco) are found in Ponsich (1988) and Aranegui (2008).

We find ourselves in the Roman city of Baelo Claudia in the proximity of the Strait of Gibraltar. The year is 195 BC and the bluefin tuna fishing season is well underway. The tuna spawning migration has long been known and the Romans have set up a fish factory here to catch tunas as they enter the Mediterranean Sea.

Everything is ready at the factory, and on the beach the fishermen await the moment. Soon afterwards, from the cape (Fig. 4.8) *a look-out signals to them that it is time to put out the nets to intercept the approaching tunas* (Fig. 4.9).

After several hours of work close to two hundred tunas have been caught, some of them with a weight of over 600 kg (Fig. 4.10).

At the end of the fishing operation the tunas are taken to the factory. Some of the meat from the tunas will be destined for the local population of Baelo Claudia, and the rest will be salted. The guts will be left to macerate and ferment in the sun in the factory to make garum. Garum was a gastronomic delicacy due to its organoleptic properties and it commanded a high price in the market. According to Plinio (23–79 AD), the historian and scientist, garum had a value comparable with the most expensive perfume.

Fig. 4.8 Sighting of a bluefin tuna bank from the cape of *Baelo Claudia* An interpretation by Cort (2006, 2007) *Artist*: Lineke Zubieta (Santander, Spain) (Documentary file, *IEO*)

Fig. 4.9 Bluefin tuna fishing in the Roman city of *Baelo Claudia*, 2nd century B.C. An interpretation by Cort (2006, 2007) *Artist*: Lineke Zubieta (Santander, Spain) (Documentary archive, *IEO*)

Fig. 4.10 Successful bluefin tuna fishing operation in *Baelo Claudia* An interpretation by Cort (2006, 2007) *Artist*: Lineke Zubieta (Santander, Spain)

4.3 Modern Age

Regarding the Modern Age there are many testimonies to the traps (Fig. 4.11) being a privilege awarded by the kings to the nobility (García 2012). In 1379, King John I recognized the property of all the traps in favor of the Count of Niebla. In 1396, Henry III similarly recognized such property as patrimony of the count's inheritance. In 1445, King John II declared the first Duke of Medina Sidonia as the owner of all space suitable for the setting up of the traps. Centuries later, a document written by the Benedictine father Friar Martín de Sarmiento (Fig. 4.12) addressed to the Duke of Medina Sidonia (López and Ruiz 2005, 2012), then owner of the traps, on the 18th of February 1757, stated that: *This observation that I have been witness to excited the curiosity of the Duke upon comparing it with the scarcity of tunas now being caught in the traps; stating that in past centuries the tunas caught in the traps were almost infinite*. When father Sarmiento wrote the phrase: ...*stating that in past centuries the tunas caught in the traps were almost infinite*, he was referring to the historical catches of 1555–1570 in which practically two traps (Conil and Zahara) caught a mean of 58,000 tunas per fishing season, though according to several contemporary scientists (J. M. Fromentin; D. Florido del Corral, J. A, López) this figure would have also included small tunas. Twenty years after those historic events catches hardly reached 5,000 tunas per trap. In the last quarter of the XVIth century the absence of tunas

Fig. 4.11 Bluefin tuna fishing with trap in the Strait of Gibraltar, around 1550. Taken from Di Natale (2012)

coincided with the desertion of the market by buyers, which led to the downfall of the traps.

In de Buen (1925) the catch statistics are presented from three Spanish traps installed in the Strait of Gibraltar between 1525 and 1756, in which maximums and minimums are shown between 1555–1570 and 1590–1756 respectively (Fig. 4.13). Studying this historical series, Gancedo et al. (2009) concluded that the low temperatures recorded between 1640–1715 during the so-called *Little Ice Age* may have reduced recruitment and the abundance of this species in the North Atlantic and Mediterranean. Sella (1929), cited by Manfrin et al. (2012), was the first to describe the existence of periodic fluctuations of 110 years in the fishing statistics from four traps of the Strait of Gibraltar and western Mediterranean Sea between 1770 and 1925. Later, Fromentin (2002) studied variability in the catches of eight traps in the eastern Atlantic and western Mediterranean between 1599 and 1960 and found periodic fluctuations of 100–120 years, which Ravier and Fromentin (2004) determined to be inversely related to sea temperature, which may have led to variations in ABFT spawning migratory patterns in response to changes in oceanic conditions.

4.4 Recent History

Until a little earlier than the middle of the XXth century the most important ABFT catches in the eastern part of the Atlantic ocean and Mediterranean sea were made

Fig. 4.12 Father Martín de Sarmiento (1695–1772). (Taken from García 2012)

mainly by traps, and it was from then onwards that other fishing systems began to be used, such as purse seine, longline and bait boat (ICCAT 2010). This new era created a bonanza situation that hardly lasted 10 years, because from 1963 there came a sharp fall in catches in the ABFT spawner fisheries of the eastern Atlantic, the Strait de Gibraltar and northern European fisheries (Fromentin and Powers 2005), so much so that the traps have since failed to return to the production levels they had reached during the first half of the last century (Cort et al. 2012) and the ABFT fisheries of the north of Europe collapsed at the beginning of the 1980s (Nøttestad and Graham 2004; Bennema 2018). In the same way, the groups of small spawners (up to 2 m), which had been abundant in the Bay of Biscay, were no longer so after 1973 (Cort and Nøttestad 2007). In a study recently published (Cort and Abaunza 2015) the relationship is studied between the high fishing mortality exerted on juveniles (<5 years) in the eastern Atlantic juvenile fisheries from 1949 to 2006, with the decline in the catch of ABFT spawners. At the start of this crisis there was no international organization dedicated to monitoring and management over tuna fisheries resources in the Atlantic and Mediterranean, since ICCAT was founded in 1966.

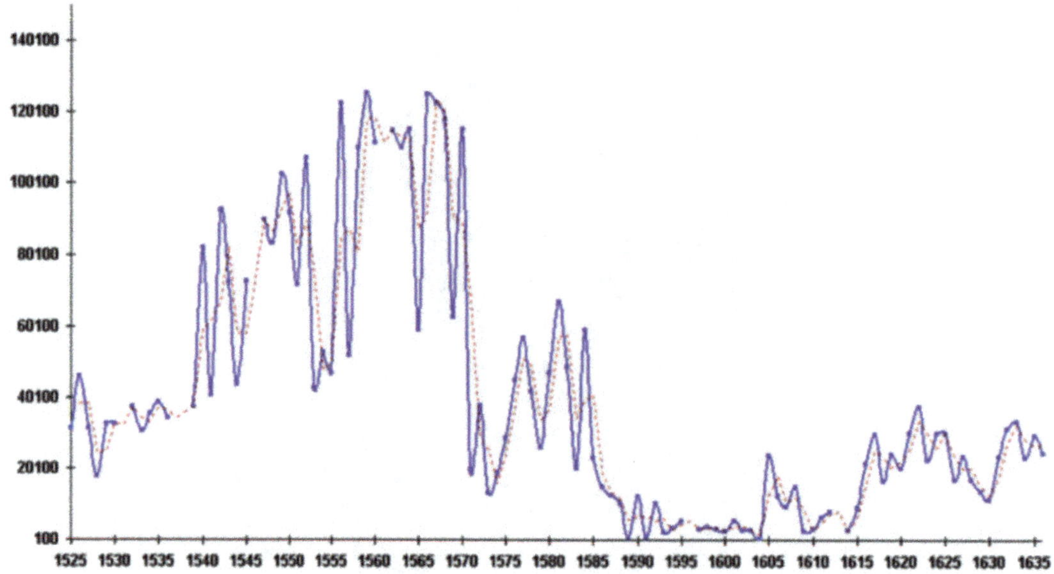

Fig. 4.13 Tuna catches (in number of fishes) of three Spanish traps (Conil, Zahara and Conilejo) between 1525 and 1635 (The dashed line is the moving mean for a period of two years)

The history of the eastern ABFT fishing in the last seventy years is plagued with events that have shaped the destiny of this species. The first of these was the collapse of the northern European fisheries at the beginning of the 1980s and the sharp fall in the catches of the traps which brought with the disappearance of the Tuna Trap Fishing National Consortium (1928–1971) ten years later (Ríos 2007; López and Ruíz 2012; Florido del Corral 2013; Florido del Corral et al. 2018). Whereas Atlantic traps fisheries survived from the 1980s thanks to the entrance of the Japanese market into the sector, pressure on stocks by the purse seine fleets rose alarmingly in the Mediterranean from the 1970s leading, thirty years later, to the greatest crisis in the history of this fishery, and in turn forcing the adoption of the PARP by ICCAT in 2007. The rigorous monitoring of the fishery since then, the implementation of severe conservation measures and highly restrictive fishing quotas, which have led to the disappearance of a large part of the fleet as well as an increase in scientific activities (Di Natale et al. 2017), including aquaculture (De la Gándara et al. 2016), have turned this fishery into a world-class example of population recovery, as described in the latest reports of the scientific committee of ICCAT, the SCRS (ICCAT 2017).

References

Allen JRM, Brandt U, Brauer A, Hubberten H-W, Huntley B, Keller J, Kraml M, Mackensen A, Mingram J, Negendank JFW, Nowaczyk NR, Oberhänsli H, Watts WA, Zolitschka B (1999) Rapid environmental changes in southern Europe during the last glacial period. Nature 400:740–743

Aranegui C (2008) Introducción a la arqueología de Lixus (Larache, Marruecos). Biblioteca virtual Miguel de Cervantes, 19 pp. http://www.cervantesvirtual.com/obra/introduccin-a-la-arqueologa-de-lixus—larache-marruecos-0/

Arévalo A, Bernal D (2007) Las Cetariae de Baelo Claudia. Avance de las investigaciones arqueológicas en el barrio meridional (2000–2004). Alicia Arévalo y Darío Bernal (Editores científicos). Junta de Andalucía. Consejería de cultura. Universidad de Cádiz, Servicio de publicaciones, 571 pp. ISBN 978-84-9828-155-2

Bennema FP (2018) Long-term occurrence of Atlantic bluefin tuna *Thunnus thynnus* in the North Sea: contributions of non-fishery data to population studies. Fish Res 199:177–185. https://doi.org/10.1016/j.fishres.2017.11.019

Cort JL (2006) El cimarrón del Atlántico Norte y Mediterráneo. Sexto concurso nacional de Ciencia en Acción. Modalidad, *Medioambiente*. Cort JL (coordinador). Instituto Español de Oceanografía. Depósito legal: SA. 494-2006, 80 pp.

Cort JL (2007) El enigma del atún rojo reproductor del Atlántico nororiental. Modalidad, *Sostenibilidad*. Octavo concurso nacional de Ciencia en Acción. José L. Cort (coordinador). Instituto Español de Oceanografía. Depósito legal: SA. 538-2007, 62 pp.

Cort JL, Abaunza P (2015) The fall of tuna traps and collapse of the Atlantic Bluefin Tuna, *Thunnus thynnus* (L.), fisheries of Northern Europe in the 1960s. Rev Fish Sci Aquac 23(4):346–373. http://dx.doi.org/10.1080/23308249.2015.1079166

Cort JL, de la Serna JM, Velasco M (2012) El peso medio del atún rojo (*Thunnus thynnus*) capturado por las almadrabas del sur de España entre 1914–2010. Col Vol Sci Pap ICCAT 67:231–241

Cort JL, Nøttestad L (2007) Fisheries of bluefin tuna (*Thunnus thynnus*) spawners in the Northeast Atlantic. Col Vol Sci Pap ICCAT 60:1328–1344

De Buen F (1925) Biología del atún, *Orcinus thynnus* (L.). Resultado de las campañas realizadas por acuerdos internacionales, 1. Madrid, 118 pp.

De la Gándara F, Ortega A, Buentello A (2016) Tuna Aquaculture in Europe. In: Benetti DD, Partridge GJ, Buentello A (eds) Advances in tuna aquaculture. From hatchery to market. Academic Press is an imprint of Elsevier. The Boulevard, Langford Lane, Kidlington, Oxford OX5 1 GB 225 Wyman Street, Waltham MA 02451, pp 115–184

Di Natale (2012) New data on the historical distribution of bluefin tuna (*Thunnus thynnus*, L.) in the Artic Ocean. Col Vol Sci Pap ICCAT 68(1):102–114

Di Natale A, Tensek S, Pagá García A (2017) ICCAT Atlantic-wide research programme for bluefin tuna (GBYP): activity report for the last part of phase 5 and first part of phase 6 (2015–2016). Col Vol Sci Pap ICCAT 73(7):2424–2503

Fernández Gómez F, Yáñez Polo MA, Hurtado Rodríguez L (2007) Surcando el tiempo. A la caza del atún rojo en las almadrabas atlánticas del estrecho. Desde la antigüedad hasta nuestros días. Fluidmecanica Sur, Club UNESCO Sevilla y Tanger. Proyecto Oceanus, 182 pp. ISBN 978-84-606-4458-3

Finlayson C, Giles Pacheco F, Rodríguez-Vidal J, Fa JDA, Gutierrez López JM et al (2006) Late survival of Neanderthals at the southernmost extreme of Europe. Nature 443(7113):850–853

Florido del Corral D (2013) Las almadrabas andaluzas bajo el consorcio nacional almadrabero (1928–1971): aspectos socio-culturales y políticos. Semana, Ciencias Sociais e Humanidades 25:117–151. ISSN 1137-9669

Florido del Corral D, Santos A, Ruiz JM, López JA (2018) Las almadrabas suratlánticas andaluzas. Historia, tradición y patrimonio (siglos XVIII–XXI). Editorial Universidad se Sevilla, 328 pp. ISBN 978-84-472-1885-1

Fromentin JM (2002) Final Report of STROMBOLI-EU-DG XIV Project 99/022. European Community-DG XIV, Brussels, 109 pp.

Fromentin JM, Powers J (2005) Atlantic bluefin tuna: population dynamics, ecology, fisheries and management. Fish Fish 6:281–306

Gancedo U, Zorita E, Solari AP, Chust G, Santana del Pino A, Polanco J, Castro JJ (2009) What drove tuna catches between 1525 and 1756 in southern Europe? International Council for the Exploration of the Sea. Published by Oxford Journals: 1595–1604

García F (2012) Las almadrabas de la costa andaluza bajo el dominio de la casa ducal de Medina Sidonia. Su tipología, sus producciones y sus problemáticas. Col Vol Sci Pap ICCAT 67:75–87

García Vargas E, Bernal D (2009) Roma y la producción de garum y salsamenta en la costa meridional de Hispania. Estado actual de la investigación. In Arqueología de la pesca en el estrecho de Gibraltar. De la prehistoria al fin del mundo antiguo. D. Bernal Casasola (Editor científico). Monografías del Proyecto Sagena 1. Servicio de publicaciones de la Universidad de Cádiz, pp 133–182. ISBN 978-84-9828-234-4

García Vargas E, Florido del Corral D (2011) Tipos, origen y desarrollo histórico de las almadrabas antiguas. Desde época romana al imperio bizantino. In: *Pescar con Arte. Fenicios y romanos en el origen de los aparejos andaluces*. D. Bernal Casasola (Editor científico). Servicio de publicaciones de la Universidad de Cádiz, pp 231–254. ISBN 978-84-9828-365-5

Giles-Pacheco F, Finalyson C, Gutiérrez JM, Santiago A, Finalyson G, Reinoso C, Giles Guzmán F (2001) Investigaciones arqueológicas en Gorham´s cave (Gibraltar): Resultados preliminares de las campañas de 1977 a 1999. The Gibraltar Caves Project. Ethel Allul/Laboratorio de prehistoria. Universidad Rovira i Virgili, Tarragona. Almoraima, 16 pp.

ICCAT (2010) ICCAT Manual. Description of species. Chapter 2; 2.1.5 Atlantic Bluefin Tuna 99:93–111. Madrid, ICCAT. http://iccat.int/Documents/SCRS/Manual/CH2/2_1_5_BFT_ENG.pdf

ICCAT (2017) Report of the 2017 ICCAT bluefin stock assessment meeting. Madrid, Spain, 22–27 July 2017, 106 pp. http://iccat.int/Documents/Meetings/Docs/2017_BFT_ASS_REP_ENG.pdf

López JA, Ruiz JM (2005) Series históricas de las capturas de atún rojo en las almadrabas del golfo de Cádiz. In Acuicultura, Pesca y Marisqueo en el golfo de Cádiz. Junta de Andalucía, Consejería de Agricultura y Pesca, pp 308–364. http://www.juntadeandalucia.es/servicios/publicaciones/detalle/48581.html

López JA, Ruíz JM (2012) Series históricas de las capturas de atún rojo en las almadrabas del golfo de Cádiz (siglos XVI–XXI). Col Vol Sci Pap ICCAT 67:139–174

Manfrin G, Mangano A, Piccinetti C, Piccinetti R (2012) Les données sur la capture des thons par les madragues dans l´archive du prof. Sella. Col Vol Sci Pap ICCAT 67:106–111

Montero J, Bernal D (2011) El estrecho de Gibraltar en el Pleistoceno. In *Benzú y los orígenes de Ceuta*. In: Ramos J, Bernal D, Cabral A, Vijante E, Cantillo JJ (coordinadores). Ciudad autónoma de Ceuta, Consejería de educación, cultura y mujer. Museo de la Basílica Tardorromana de Ceuta, Universidad de Cadiz, pp 130–131. ISBN 978-84-15243-14-4

Morales A, Roselló E (2007) Los atunes de *Baelo Claudia* y Punta Caraminal. In Las Cetariae de Baelo Claudia. Avance de las investigaciones arqueológicas en el barrio meridional (2000–2004). Alicia Arévalo y Darío Bernal (Editores científicos). Junta de Andalucía. Consejería de cultura. Universidad de Cádiz, servicio de publicaciones, pp 489–498. ISBN 978-84-9828-155-2

Niveau de Villedary AM (2009) Ofrendas, banquetes y libaciones. El ritual funerario en la necrópolis púnica de Cádiz. Spal Monografías XII. Servivio de publicaciones, Universidad de Cádiz (ISBN 978-84-9828-258-0). Secretariado de publicaciones, Universidad de Sevilla (ISBN 978-84-472-1203-3), 297 pp.

Niveau de Villedary AM (2011) Algunos indicios sobre la (posible) práctica de sacrificios humanos en Cádiz. In *Cultos y ritos en la Gadir fenicia*. Servivio de publicaciones, Universidad de Cádiz (ISBN 978-84-9828-337-2). Secretariado de publicaciones, Universidad de Sevilla (ISBN 978-84-472-1343-6), pp 405–420

Nøttestad L, Graham N (2004) Preliminary overview of the Norwegian fishery and science on Atlantic bluefin tuna (*Thunnus thynnus*). Scientific report from Norway to ICCAT Commission meeting in New Orleans, USA, 15–21 Nov 2004, 12 pp.

Ponsich M (1988) Aceite de oliva y salazones de pescado. Factores geo-económicos de Bética y Tingitania. Editorial Universidad Complutense, 253 pp. ISBN-84-7491-248-2

Ravier C, Fromentin JM (2004) Are long-term fluctuations in Atlantic bluefin tuna (*Thunnus thynnus*) populations related to environmental changes? Fish Oceanogr 13:145–160

Ríos Jiménez S (2007) La gran empresa almadrabero conservera andaluza entre 1919 y 1936: el nacimiento del Consorcio Nacional Almadrabero. Historia Agraria 41:57–82

Rodríguez-Vidal J, Finlayson G, Finlayson C, Negro JJ, Cáceres LM, Fa DA, Carrión JS (2013) Undrowning a lost word—the marine isotope stage 3 landscape of Gibraltar. Geomorphology 203:105–114. http://www.elsevier.com/locate/geomorph

Sarà R (1998) Dal mito all'aliscafo. Storie di Tonni e Tonnare. Banca Aegusea Ed., Favignana-Palermo, pp 1–271

Sella M (1929) Migrazioni e habitat del tonno (*Thunnus thynnus*) studiati col metodo degli ami, con osservazioni ull'accrescimento, sul regime delle tonnare, ecc. Memorie, R. Comitato Talassografico Italiano 159:1–24

Stringer CB, Finlayson JC, Barton RNE, Fernández-Jalbo Y, Cáceres I, Sabin RC, Rhodes EJ, Currant AP, Rodríguez-Vidal J, Giles-Pacheco F, Riquelme-Cantal JA (2008) Neanderthal exploitation of marine mammals in Gibraltar. Proc Natl Acad Sci 105(38):14319–14324

Tzedakis P, Lawson IT, Frogley MR, Hewit GM, Preece RC (2002) Buffered tree population changes in a quaternary refugium: evolutionary implications. Science 297:2044–2047

Chapter 5
The Present State of Traps and Fisheries Research in the Strait of Gibraltar

Abstract The Strait of Gibraltar is defined as one of the most emblematic places in bluefin tuna biology, and the traps installed therein are platforms without equal in the study of this species. The enormous literature available on his fishing gear makes it impossible to make a full description from the historical point of view in the present study; however, some of the scientific activities and scientific committees of the last century are cited, followed by a list of the most recent research projects on the species. Two symposia, held in 2008 and 2011, dealing with bluefin tuna biology and fishing are reported, the latter of which was exclusively dedicated to the traps fishery and the research carried out at these fishing installations.

The Strait of Gibraltar (Fig. 5.1) is one of the most emblematic sites of the ABFT life cycle. Eastern Atlantic spawners pass through it in the boreal spring (April–June) towards the Mediterranean spawning grounds (Frade 1938; Vilela and Cadima 1961; Rodríguez-Roda 1964; Duclerc et al. 1973; Piccinetti et al. 1997; Tsuji et al. 1997; Medina et al. 2002; García et al. 2006; Suzuki and Kai 2012) and a few weeks later they swim back to the Atlantic Ocean (Rodríguez-Roda 1969a, b; Mather et al. 1995; Block et al. 2005; Medina et al. 2011; De la Serna et al. 2013; Aranda et al. 2013; Quílez-Badia et al. 2013) where they spend the rest of the year feeding intensively as far as the northernmost waters of the North Atlantic (Hamre 1960; Hamre 1965; ICCAT 1996; MacKenzie et al. 2014) at 75° N–1° E (De Metrio et al. 2002), or even to the western side of the Atlantic (Block et al. 2005). A part of the juveniles born at the end of spring in the western Mediterranean pass through the strait at the beginning of autumn (October), according to Rey (1979) and Rey and Cort (1986), and concentrate in wintering areas to the south of Morocco (Lamboeuf 1975; Brêthes 1978, 1979; Brêthes and Mason 1979; Rey and Cort 1986; Cort 1990).

Since ancient times ABFT has been caught using traps, a fishing gear that at first consisted of nets thrown from land, also known as traps de *vista* or *tiro* (Sañez 1791; Florido del Corral et al. 2018), as illustrated in previous pages; but nowadays traps de *buche* are used (Florido del Corral et al. 2018), which are made up of a complex system of chambers through which the tunas, after being guided in by kilometric long nets called *raberas*, the fishermen pass them from one to another until the last, or *chamber of death* where they are finally killed (Fig. 5.2).

J. L. Cort and P. Abaunza, *The Bluefin Tuna Fishery in the Bay of Biscay*, SpringerBriefs in Biology, https://doi.org/10.1007/978-3-030-11545-6_5

Fig. 5.1 The Strait of Gibraltar (https://en.wikipedia.org/wiki/Strait_of_Gibraltar#/media/File:STS059-238-074_Strait_of_Gibraltar.jpg)

On the web page of the Organization of Fisheries Producers[1] it says that the *buche traps* come from the Mediterranean and that they became consolidated on the Cádiz coasts in the last third of the nineteenth century. They describe it thus:

> They are located approximately three kilometers from the coast and are more or less 34 meters deep, depending on each trap. The trap presents a complicated structure made up of a large skeleton of cables on which the nets rest, held to the bottom by means of lead weights and chains, and sustained by corks or floats in the upper part. It can be divided into two essential parts: the Catcher, mainly made up of the body of trap and the Auxiliary, which is the leader from shore to trap and the leader from offshore to trap.

ABFT fishing with traps in the Strait of Gibraltar has been the most commonly used system over the long history of this fishery (Sáñez 1791; De Buen 1925). In this

[1] http://fis.com/fis/companies/details.asp.

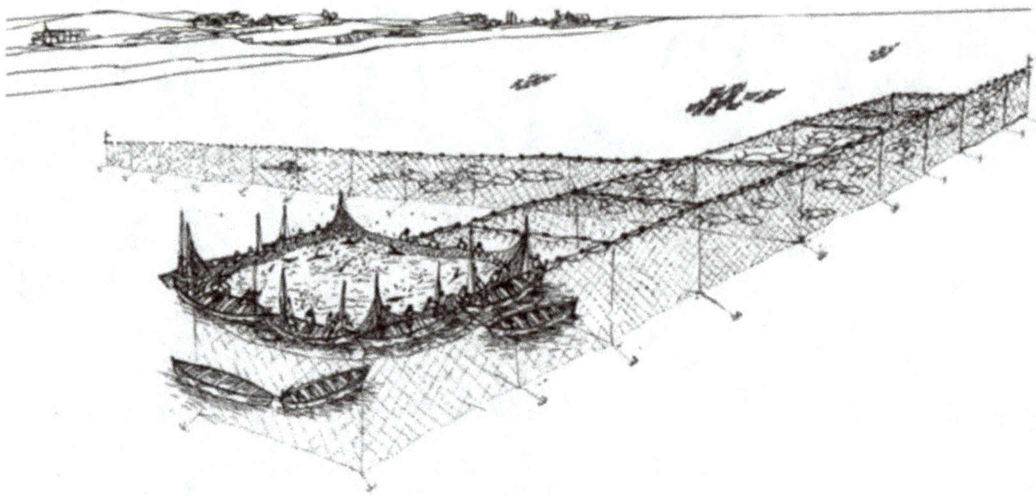

Fig. 5.2 Trap still in use (http://www.photolib.noaa.gov/bigs/fish2059.jpg)

Fig. 5.3 Bluefin tuna in the spawning phase caught in the trap of Barbate, 2009 (Cádiz, Spain) (Documentary file, *IEO*)

area it has traditionally been used to catch spawners (Fig. 5.3) both in the spawning and the trophic phases (Lozano 1958; Rodríguez-Roda 1964, 1980; Pereira 2012).

Sella (1929), cited by Manfrin et al. (2012) first presented the existence of periodic fluctuations of 110 years in the fishing statistics of four traps of the Strait of Gibraltar and western Mediterranean Sea between 1770 and 1925. Rodríguez-Roda (1978) found periodic fluctuations in the catches of the traps in the south of Spain between 1929 and 1977 with maximums each 6 or 7 years. Later, in 2002, variability in the catches of eight traps in the eastern Atlantic and western Mediterranean between 1599 and 1960 was presented, revealing periodic fluctuations of 100–120 years that were inversely related to sea temperature, which may have led to variations in the migratory patterns of ABFT spawners in response to changes in oceanographic conditions (Frometin 2002).

The first research studies on tunas conducted by scientists of the *Instituto Español de Oceanografía* (*IEO*) are included in a compendium of publications entitled: "Resultados de las campañas realizadas por acuerdos internacionales" (*Results of the campaigns carried out under international agreements*), published between 1925 and 1927 (De Buen 1925), directed by Professor Odón De Buen (Fig. 5.4). The book is made up of several articles on the biology and fishing of bluefin tuna, *Orcynus thynnus* (L.) in the south of Spain signed by Fernando De Buen, then head of the Department of Biology at the *Dirección General de Pesca* (General Directorate of Fisheries); Luis Bellón Uriarte, assistant to the *IEO*, and Álvaro de Miranda y Rivera, the head of the Oceanographic Laboratory of Málaga (*IEO*).

These studies beautifully define the fishing activity of the traps; the landing statistics of the time and before, even those of Father Sarmiento (1525–1750), and of the tuna canning industry. All of this was accompanied by splendid illustrations by Luis Bellón as well as photographs of fishing in the traps and tuna processing at factories (Figs. 5.5 and 5.6).

> According to Florido del Corral (2013): The establishment of Tuna Trap Fishing National Consortium (1928–1971) meant a commitment by the State, of a transformation of the enterprise and social organization of the Fishing Industry at the Andalusian Atlantic coasts (SW, Spain), confirming processes previously announced and inaugurating new ones. Business concentration, productive rationalization of manufacturing sites, and the consolidation of the company-towns crystallized in the central decades of 20th century, transforming traditional elements of the work cultures of the tuna fisheries, such as forms of retribution. These processes are better understood if we analyse them contextualised in the frame of power networks of the tuna fisheries oligarchy, which were characteristic in Andalusian and Spanish societies. From this holistic perspective, it is concluded the conflicting nature of the evolution of the tuna industry, between tradition and the productive rationalization.

While the Tuna Trap Fishing National Consortium was established (Ríos 2007; Florido del Corral 2013), a great deal of information was accumulated on the fishing activities and landing statistics Spanish traps, which were of great scientific value. Spanish scientists enjoyed the benefit of these advances to make their contribution to improving knowledge of tunas, mainly of ABFT.

The Spanish Civil War and the Second World War stalled the progress of non-war related scientific activities, but the end of these conflicts saw the beginning of great changes resulting from the need to find food originating from the sea for a population in great need.

In the 1950s studies of great interest were published such as that of Julio Rodríguez-Roda (*Instituto de Investigaciones Pesqueras*of *CSIC*, Cádiz), on the ABFT of the Strait of Gibraltar (Rodríguez-Roda 1957), and a voluminous study

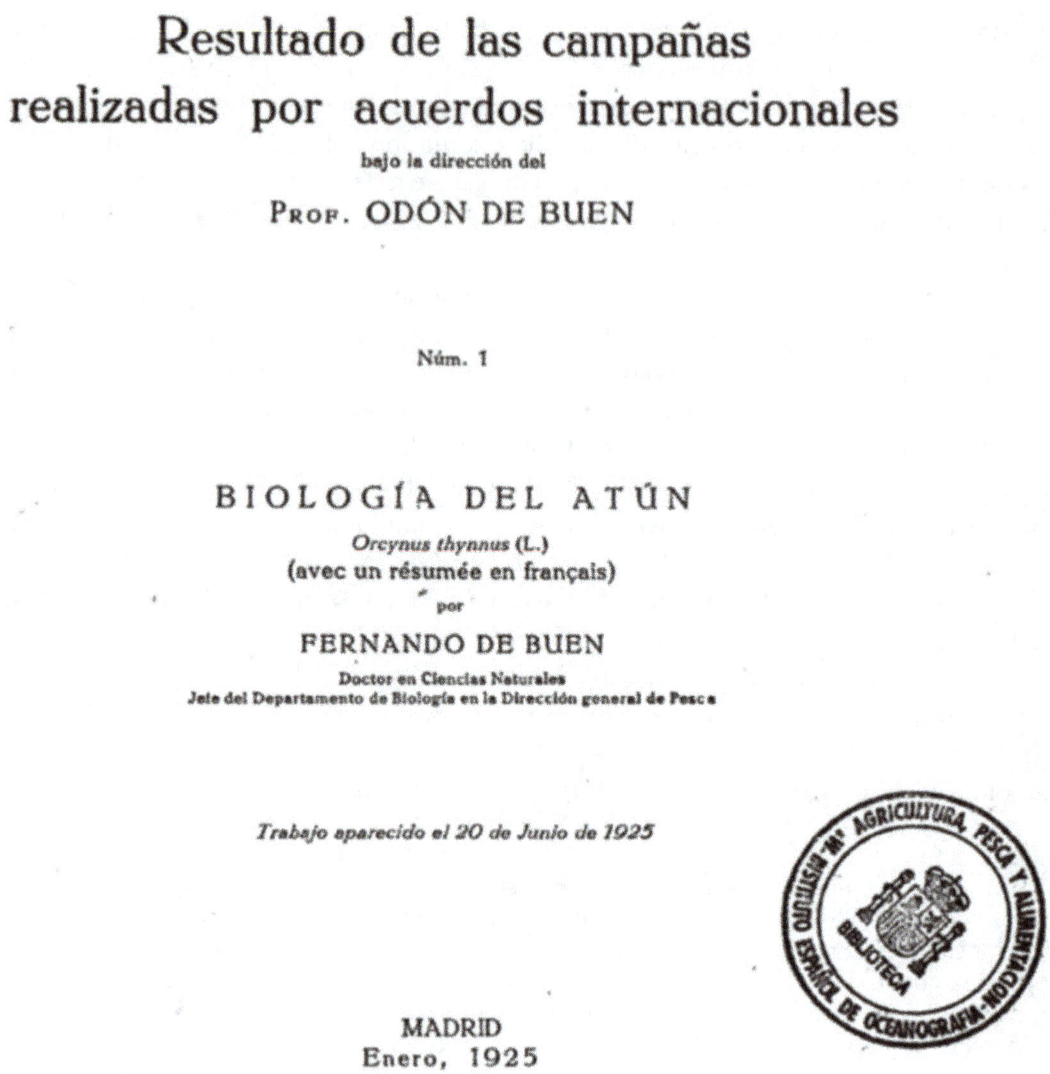

Resultado de las campañas
realizadas por acuerdos internacionales
bajo la dirección del
PROF. ODÓN DE BUEN

Núm. 1

BIOLOGÍA DEL ATÚN
Orcynus thynnus (L.)
(avec un résumée en français)
por
FERNANDO DE BUEN
Doctor en Ciencias Naturales
Jefe del Departamento de Biología en la Dirección general de Pesca

Trabajo aparecido el 20 de Junio de 1925

MADRID
Enero, 1925

Fig. 5.4 First studies of tunas at the IEO

of the Scombrids of Spanish and Moroccan waters by Fernando Lozano Cabo (*IEO*), published in 1958 (Lozano 1958; Fig. 5.7).

International committees and working groups of the *International Council for the Exploration of the Sea* (*ICES*) already existed in those years (Hamre and Tiews 1963) and of the *General Fisheries Commission for the Mediterranean* (*GFCM*) (Lozano 1959) in which ABFT was studied by eminent scientists, such as Massimo Sella, Pascuale Arena and Raimondo Sarà (Italy), Fernando Frade and H. Vilela (Portugal), Henri Heldt (Tunisia-France), Jean Le Gall (France), Johanes Hamre (Norway), Klaus Tiews (Germany), Frank Mather III and Luis Rivas (U.S.A.), Akira Suda (Japan), and Julio Rodríguez-Roda and Fernando Lozano (Spain), among others.

Dr. Rodríguez-Roda retired at the beginning of the 1980s and since then scientists of the *IEO* have continued in their task, fulfilling the commitments acquired by Spain in the scientific research into this fishery. At first these were Juan C. Rey and Juan A. Camiñas, and later on José M. de la Serna, J. Ortiz de Urbina and D. Macías.

The work Rodríguez-Roda left was immense and is nowadays still fundamental to any study into the biology and dynamic of ABFT. To mention some of the studies, Figs. 5.8 and 5.9 are examples of two publications on the biology of the ABFT, but he also worked on the fecundity, growth, ethology and environment of the species and conscientiously followed ABFT catches in the traps of southern Spain from the point of view of their biology and fishing yield (Rodríguez-Roda 1964).

After the great advance on the knowledge of the ABFT carried out during the first half of the 20th century, in recent decades there has been a great deal of scientific production contributing considerably to our knowledge of the biology of ABFT and its population dynamic, precisely at a time in which the species has shown signs of fishing overexploitation, mainly in the last decades (ICCAT 2008). Below some publications are cited, listed by their general subject matters:

Reproduction and fecundity: Rodríguez-Roda (1967), Baglin (1982), Medina et al. (2002), Karakulak et al. (2004), Corriero et al. (2005), García et al. (2005), Heinish et al. (2014), Addis et al. (2016), Richardson et al. (2016).

Larvae and larval ecology: Piccinetti and Piccinetti Manfrin (1970), Dicenta and Piccinetti (1980), García et al. (2006), Alemany et al. (2010), Álvarez-Berastegui et al. (2014), Reglero et al. (2012), Laiz-Carrión et al. (2015).

Fig. 5.5 Old tuna canning factory, around 1925 (Adapted from Bellón, 1926)

Fig. 5.6 "The tuna forest" in a canning factory of tuna caught by the traps, around 1925 (Documentary archive, *IEO*)

Feeding: Estrada et al. (2005), Sarà and Sarà (2007), Logan et al. (2010), Butler et al. (2015), Battaglia et al. (2013), Sorell et al. (2017).

Ecology: Druon et al. (2011, 2016).

Natal origin: Rooker et al. (2014), Fraile et al. (2014), Brophy et al. (2015).

Genetics: Puncher et al. (2015, 2018).

Age and growth: Rodríguez-Roda (1964), Caddy et al. (1976), Butler et al. (1977), Compeán-Jiménez and Bard (1983), Cort (1990), Santamaria et al. (2003), Rodríguez-Marín et al. (2007), Restrepo et al. (2010), Cort et al. (2013, 2014, 2015), Luque et al. (2014), Ailloud et al. (2017).

Electronic tagging[2]: De Metrio et al. (2002), Block et al. (2005), Goñi et al. (2010), Wilson and Block (2009), Medina et al. (2011), Aranda et al. (2013), Lutcavage et al.

[2]The internal electronic tags (*archival tags*) are small computers that are placed in the peritoneal cavity of the ABFT, which record dates, times, fish depths, water temperature, body temperature and light levels, which are used to calculate an approximate daily position of the tagged animal depending on the times of dawn and dusk and the angle of the sun. To download the data the fish must be recovered. The electronic tags can record data every few seconds over several years depending on the tag´s sampling frequency and the duration of the battery.
The electronic *pop-up* tags, also known as *PAT satellite tags*, compile information on oceanic movements and preferred water temperature, clarity and currents through GPS location technology. They are pre-configured to come loose at a programmed time and rise to the surface to transmit the

Fig. 5.7 Bluefin tuna of 530 kg caught in the trap of Barbate (Spain). Strait of Gibraltar, 6/24/1954 Photo: F. Lozano (1958)

(2013), Quílez-Badia et al. (2013), Cermeño et al. (2015), Galuardi et al. (2015), Abascal et al. (2016), Di Natale et al. (2016), Tensek et al. (2018).

Aerial and acoustic prospection: Lutcavage et al. (1995, 1997), Vanderlaan et al. (2014), Melvin (2016), Goñi et al. (2017), Rouyer et al. (2018).

Natural mortality: Brodziak et al. (2011), Fonteneau and Maguire (2014).

Stock assessment: Anonymous (1994), Fromentin et al. (2014), ICCAT (2017).

In several published synopses (Mather III et al. 1973, 1995; Fromentin and Powers 2005; Rooker et al. 2007; ICCAT 2010) and numerous studies carried out in the

data to the Argos satellite network. This network collects, processes and disseminates environmental data and has a special channel dedicated to wildlife telemetry.

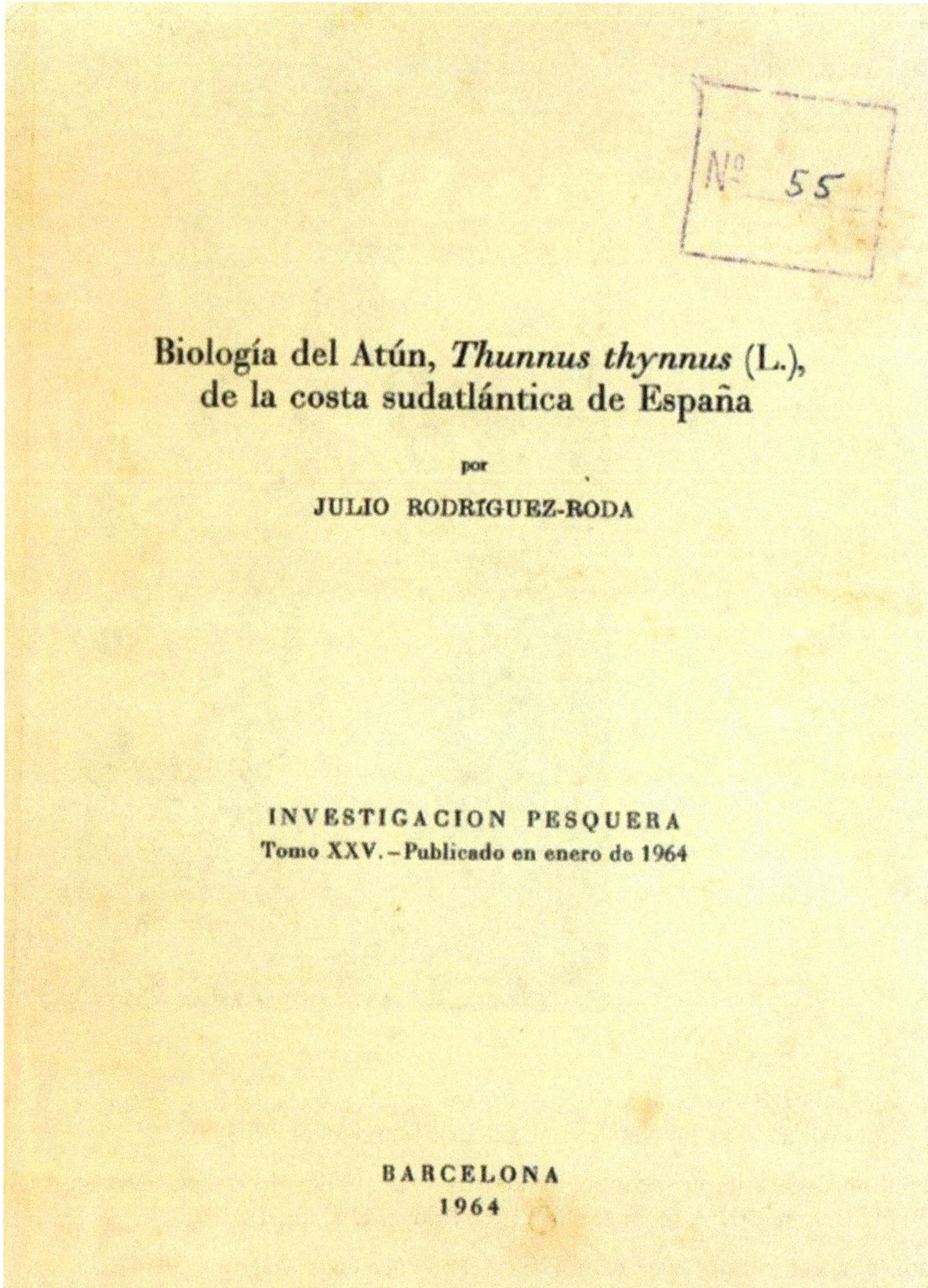

Biología del Atún, *Thunnus thynnus* (L.), de la costa sudatlántica de España

por

JULIO RODRÍGUEZ-RODA

INVESTIGACION PESQUERA
Tomo XXV.–Publicado en enero de 1964

BARCELONA
1964

Fig. 5.8 Cover of the article by Rodríguez-Roda (1964)

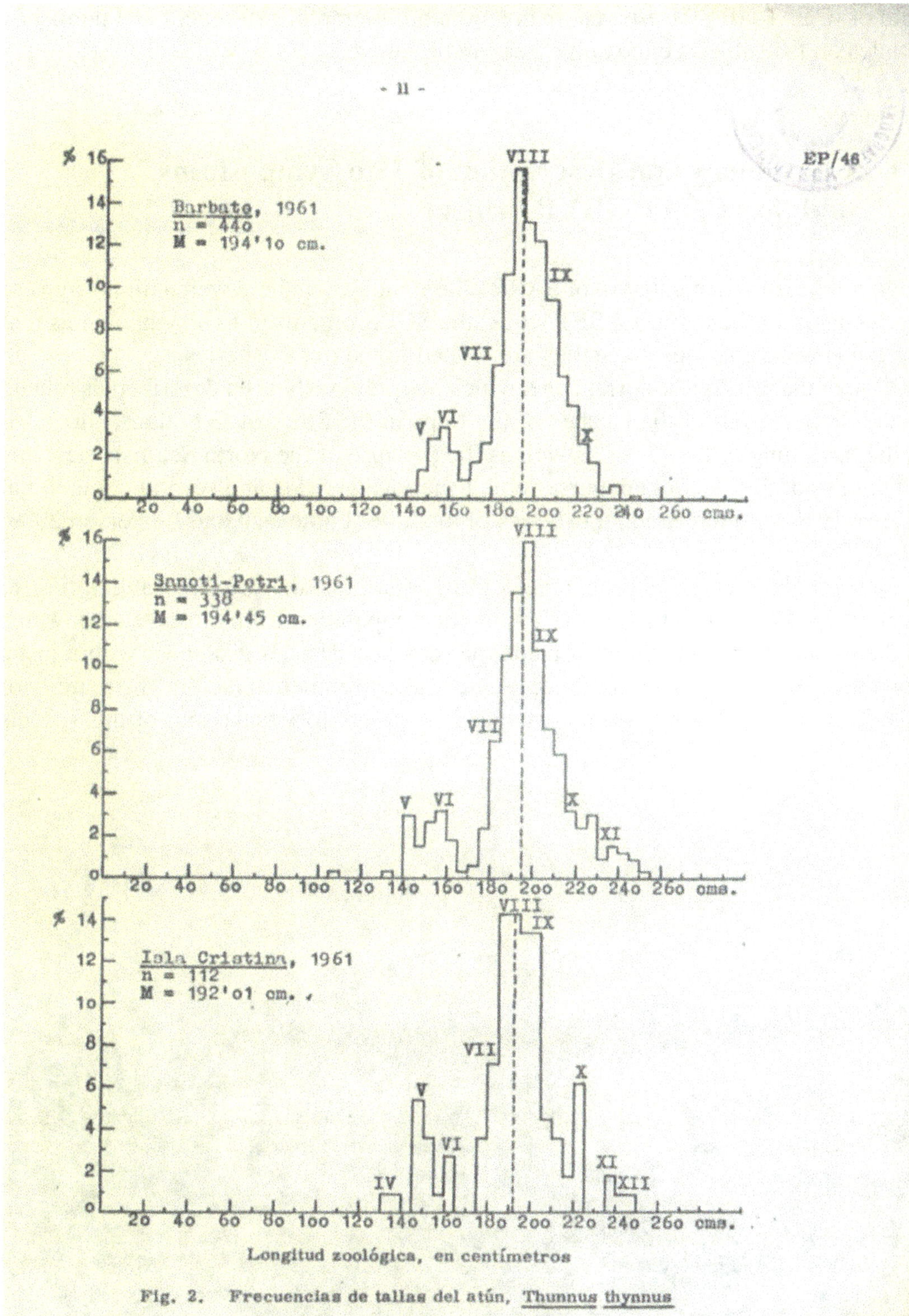

Fig. 5.9 Length distribution of bluefin tuna caught by the traps, expressing the ages of the different groups in Roman numbers, according to Rodríguez-Roda (1962)

field of ICCAT-GBYP Research, many more bibliographic references and published articles on the subjects cited and other can be found.

5.1 Chronology and Description of Two Symposiums and the ICCAT-GBYP Project

Soon after the implementation of PARP (2008) and when there were still no signs of the recovery of the eastern ABFT stock, the SCRS organized two symposiums that are going to be analysed since they are related to the traps fisheries.

One of the facts of the current age, which has previously been described, is related to the dramatic fall of the catches of the traps in the Strait of Gibraltar (Fig. 5.10) at the beginning of the 1960s as well as the collapse of the North Sea fisheries; this is the case of the Norwegian purse seine fishery (Fig. 5.11) and several commercial (Tiews 1978) and recreational fisheries of the time (Fonteneau and Le Person 2009; Ross 2010).

Several theories have been put forward regarding these events attempting to explain what happened to the ABFT spawning population from those years onwards in the eastern Atlantic, and what made the spawner fisheries disappear within just a few years. Several publications have pointed to environmental factors, the scarcity of ABFT preys in trophic areas and changes in the migratory behaviour of this species

Fig. 5.10 Trap of Sancti-Petri (Spain), 1949 (Documentary archive, *IEO*)

Fig. 5.11 Norwegian purse seiner (1957) (Courtesy of IMR, Norway)

as the main reasons (Tiews 1978; Fromentin 2002, 2009; Fromentin and Powers 2005; Ravier and Fromentin 2001, 2004; Fromentin and Restrepo 2009).

With the aim of resolving the mystery of the decline in the spawning stock, which sustained the traps fisheries and those of the north of Europe until the 1960s, Cort and Nøttestad (2007) proposed at the SCRS meeting in 2006 that a symposium be held to debate the issue. The proposal was recommended by the scientific committee and later accepted by ICCAT (2007). It took a year to organize and find sponsors, and the event was finally held in Santander (Spain) in April 2008 (Fig. 5.12). The Symposium also considered the Pacific bluefin tuna, *Thunnus orientalis* (T and S) fisheries, a species that has found itself in a similar situation to ABFT for decades. Similarly, an overall vision of the southern bluefin tuna (*Thunnus maccoyii,* C.) fishery was presented.

5.1.1 World Symposium on Bluefin Tuna in Santander (2008)

The following is a literal transcription of some of the content of the report in ICCAT (2009).

5.1.1.1 Background

The aim of the Symposium was to provide a deeper investigation of events that took place decades ago and to improve the understanding of these intriguing past events. This information should further help in improving current management and conservation measures for bluefin tuna fisheries.

The Symposium was a response to a recommendation of the Standing Committee on Research and Statistics (SCRS) in 2006.

Previous studies carried out within the framework of ICCAT have stressed the disappearance of some past fisheries or the drastic fall in the yields of others that generate changes in the spatial distribution of the catches.

PESCA

Expertos analizarán en Santander la «delicada» situación del atún rojo

En el marco de este evento se abordará la evolución histórica y las causas de la desaparición de algunas pesquerías históricas de esta especie

ALERTA / SANTANDER

Un centenar de científicos y expertos procedentes de todo el mundo participarán en el Simposio Mundial para el Estudio de la Fluctuación de los Stocks de Atún Rojo que se celebrará del 22 al 24 de abril en Santander. En el marco de este evento se analizará la evolución histórica y las causas de la desaparición de algunas pesquerías históricas de esta especie que atraviesa una situación «delicada» y «peligrosa» y se enfrenta a un futuro «muy incierto». El objetivo del Simposio, organizado por Instituto Español de Oceanografía (IEO), junto con la Comisión Internacional para la Conservación del Atún Atlántico y en colaboración con la Sociedad Regional Cantabria I+D+i (Idican), el Ministerio de Educación y Ciencia y el Ayuntamiento de Santander, es extraer conclusiones sobre la caída de pesquerías y proponer actuaciones para proteger la especie y mejorar su situación. Las proposiciones se trasladarán a la Comisión Internacional para la conservación del atún rojo. La jornada, que se desarrollará en siete sesiones temáticas relacionadas, todas ellas, con la situación presente y futura de estos túnidos, fue presentada ayer, jueves 17, en rueda de prensa por el delegado del Gobierno en Cantabria, Agustín Ibáñez, el director general de Pesca del Gobierno regional, Fernando Torrontegui, y el director del Instituto Oceanográfico de Santander, José Luis Cort. Entre las causas de la desaparición del atún rojo en pesquerías históricas, Cort destacó la sobreexplotación de la especie. Un fenómeno al que hay que sumar, según dijo, los efectos ambientales derivados del cambio climático que han repercutido en países del norte de Europa y la contaminación, que fue «uno de los argumentos principales» de la desaparición.

Agustín Ibáñez, Fernando Torrontegui y José Luis Cort en la presentación de la jornada. / DELTA

La cuota no llega a 1.200 toneladas para los barcos españoles

Durante la presentación del Simposio Mundial, Torrontegui señaló que la cuota de pesquería de atún rojo reservada en 2008 para los 63 barcos que faenan en el Cantábrico -19 de los cuales son cántabros- no llega a las 1.200 toneladas mientras que en el conjunto estatal asciende a 5.378 toneladas. La cuota total para todos los países del Atlántico es de 30.000 toneladas aunque, anunció, la pesquería en estas aguas tendrá una reducción paulatina en los próximos años. Reconoció que la pesca del atún rojo en Cantabria es «muy baja» en la actualidad a la par que calificó de «gravísimos» los «problemas» que afectan a la especie.

Fig. 5.12 Press conference on the de bluefin tuna symposium in the Spanish Government Delegation in Santander *Diario Montañés*, Santander (4/18/2008)

Although these events occurred in past decades, they have marked the future of the fisheries. In the Atlantic, these events occurred in the 1960s, whereas in the Pacific, during the late 1800s and early 1900s, several fisheries that occurred in northern Japan suddenly disappeared, while more recently, several new fisheries have started in the Sea of Japan and coastal areas of northern Japan.

The Symposium was jointly organized by ICCAT and the *Instituto Español de Oceanografía* (Spanish Institute of Oceanography), *IEO*.

The *IEO* and ICCAT jointly organized the "World Symposium for the study of fluctuations in northern bluefin tuna (*Thunnus thynnus and Thunnus orientalis*), including historical periods". At the event, which was held in Santander (Spain) between 22nd and 24th April 2008, 85 scientists from all over the world took part.

In 2007, the president of the SCRS, named a permanent committee made up of doctors J. M. Fromentin (France), J. Powers (U.S.A.) and N. Miyabe (Japan); this committee was coordinated by J. L. Cort (Spain).

The Symposium opened on April 22, 2008 with an official opening ceremony presided by Dr. Fabio Hazin, ICCAT Chair. Dr. Hazin thanked the Government of Cantabria and the city of Santander for hosting the meeting.

The ICCAT Chair emphasized the opportunity the Symposium presented at a time when the stock of North Atlantic bluefin tuna, particularly in the eastern Atlantic and Mediterranean Sea, was facing one of its worst crises in the history of the fishery. Dr. Hazin expressed the wish that the Symposium would help the SCRS to better assess the bluefin tuna stock and therefore contribute towards improving the management of the stocks.

Other scientific authorities of national and local administrations, as well as the municipal authority, spoke before the debate was opened.

The Symposium was organized in seven topical sessions coordinated by a moderator.

The present study only considers the sessions dedicated to *Thunnus thynnus*.

The documents of the symposium are available at ICCAT (2009).

5.1.1.2 Session 1: Historical Synthesis of the Bluefin Tuna Fisheries

SCRS/20108/058. *Fonteneau, A.* Atlantic bluefin tuna: An overview of 100 centuries of moving fisheries.
SCRS/2008/059. *Fonteneau, A. and A. Le Person*. Bluefin fishing in Lannion Bay, northern Brittany, during the 1946–1953 period.

Moderador: G. Scott (NOAA, U.S.A.)
Speaker: A. Fonteneau (IRD, France)

- Bluefin tuna fisheries have been identified in different areas of the Atlantic and Mediterranean over the last 10,000 years. Since the beginning of the XXth century these fisheries have undergone considerable industrial development.
- The historical analysis confirms large-scale migrations and shows that bluefin tuna has been caught in a wide range of ecosystems between 2 and 29 °C, many changes

being observed in where and when the species is caught. This fact is difficult to explain, although it would appear to be a combination of environmental and fisheries factors; exploring these more deeply will contribute to an improvement in the quality of assessments on the state of the stocks.

- A practical example to orientate future studies regarding the previously mentioned spatial-temporal changes concerns the recreational fisheries of Northern Europe in the middle of the XXth century; such is the case of the fishery in *Trebeurden Bay* (*British Channel*), practiced between 1946 and 1953. This small fishery, associated with the presence of sardines in the area, provided a great deal of useful biological information related to the North Sea and Norwegian coastal bluefin tuna fisheries.

The Symposium noted it is apparent that there are a number of important fishery dynamics that haveoccurred which have not been taken directly into account in our recent assessments of bluefin stock status in the ICCAT Convention area and concluded from Session 1 that:

- Our assessments of Atlantic bluefin stock status mainly focus on recent history, for which more detailed information is available about the catch, effort and size composition of the catch.
- These assessments are uncertain, especially so regarding biomass levels that are necessary to meet the requirements of the ICCAT Convention.
- Incorporation of more historical information could better inform us of stock productivity and abundance levels consistent with the ICCAT Convention objectives.
- Our challenge will be to apply methods for stock status assessments that are more appropriate to the added complexities of the history we can piece together.

5.1.1.3 Session 2: The collapse of the North Sea and Norwegian Coastal Bluefin Tuna Fisheries

SCRS/2008/060. *Tangen, M*. The Norwegian fishery for Atlantic bluefin tuna.
SCRS/2008/061. *Nøttestad, L., Ø. Tangen and S. Sundby*. Norwegian fisheries since the early 1960s: What went wrong and what can we do?
SCRS/2008/062. *MacKenzie, B. R. and R. A. Myers*. The development of the northern European fishery for North Atlantic bluefin tuna, *Thunnus thynnus*, during 1900–1950.

Moderator: B. McKenzie (INRA, Denmark)
Speaker: L. Nøttestad (IMR, Norway)

- Different descriptions of the ecosystem, before, during and after the disappearance of bluefin tuna from these fisheries, suggest that the collapse was not caused by great changes in water temperature, since bluefin tuna was present in times of high temperatures until 1960, and was also later caught in places where the temperature was as low as 3 ºC.
- The coincidence of the collapse of herring *stocks* in the North Sea in the 1960s may explain the disappearance of medium-sized spawning bluefin tuna (<2 m) from

these traditional fishing areas in the same years; large adults (>2 m) were still present even with a scarcity of herrings for a few more years until their complete disappearance at the beginning of the 1980s.
The following causes are suggested that may have led to the absence of bluefin tuna in these fisheries:

- Changes in migratory behavior,
- Scarcity of adults in the population, and/or
- The high exploitation of juveniles in other areas (Bay of Biscay, coasts of Portugal and Atlantic fisheries of Morocco).

- On the other hand, the ideal conditions of the ecosystem since the 1990s with a great abundance of bluefin tuna prey including herring, mackerel and even anchovy, and optimal sea temperatures, leads to the conclusion that the fundamental reason behind the bluefin tuna's disappearance and failure to return to these latitudes is the precarious condition of the population in the ocean.

The reasons why these fisheries are difficult to reconstruct are:

- The continual fall in spawning biomass over the last 10–12 years, which at its lowest point since the first records of ICCAT. The quantity of older fishes in the population has fallen,
- The increase in the exploitation of juveniles since 1950. With an exploitation pattern in which 80–90% of fishes caught are aged 1–3 years (<30 kg), together with undeclared catches of age 0 fishes (<2 kg), a large part of the bluefin tunas recruited will never have the chance to reproduce.

The Symposium raised a number of research questions in the discussion of presentations in Session 2:

- The role of learning of migration patterns by young tuna from older tuna, and the necessity for overlap of spatial distributions of young and old tuna; the mechanisms by which learning is accomplished are unclear;
- The possibility to acquire observational and occasional landings data via commercial and sport fishermen targeting other species and via whaling observers in northern European waters;
- Possible links via migration to the population in the West Atlantic;
- Attempts to estimate biomass in the early 1950s from age composition of catches and cohort identification;
- Uncertainty of stock-recruit relationship;
- The potential for individual tuna to skip spawning and subsequently to reduce fidelity to former spawning sites; and
- The role of squid abundance and prey (especially herring) condition on bluefin tuna diets and condition.

5.1.1.4 Session 3: Decline of the Adult Fisheries of the Cantabrian Sea

SCRS/2008/063. *Cort, J. L.* The bluefin tuna (*Thunnus thynnus*) fishery in the Bay of Biscay.
SCRS/2008/064. *Cort, J. L. and E. Rodríguez-Marín*. The bluefin tuna (*Thunnus thynnus*) fishery in the Bay of Biscay. Evolution of 5+ group since 1970.
SCRS/2008/065. *Cort, J. L., P. Abaunza and G. De Metrio*. Analysis of the northeast Atlantic juvenile bluefin tuna (*Thunnus thynnus*) population between 1949 and 1960.
SCRS/2008/066. *Rodríguez-Marín, E., J. M. Ortíz de Urbina, E. Alot, J. L. Cort, J. M. de la Serna, D. Macias, C. Rodríguez-Cabello, M. Ruiz and X. Valeiras*. Following bluefin tuna cohorts from east Atlantic Spanish fisheries since the1980s.

Moderator: E. Rodríguez-Marín (IEO, Spain)
Speaker: J. L. Cort (IEO, Spain)

- The fishery is traditionally made up of juveniles (1–4 years; between 4 and 35 kg). In the past there was a constant presence of medium-sized adults (up to 2 m) in the Bay of Biscay on the trophic migration from the spawning areas of the Mediterranean to feeding grounds in the North Sea. These groups have now practically disappeared from the fishery.
- A sharp fall in the abundance of fishes aged 5+ (mean weight 62 kg) has been observed since the beginning of the 1970s in a fishery dominated by the catch of juveniles for the last three decades: 96.6% of the catch in number of fishes are juveniles.
- The results of an analysis of the population of the Atlantic juvenile fisheries between 1949 and 1960 show that under different scenarios the high fishing mortality exerted during the years studied may have contributed to the fall in the spawning population of the eastern Atlantic and as a result been one of the causes of the decline in the fisheries of the north of Europe and the traps from 1963. The hypotheses that underlie the analysis show the existence of a certain degree of independence ("resident populations") of eastern and western Mediterranean juvenile bluefin tunas based on recent studies using electronic pop-up tags.

The Symposium raised a number of research questions in the discussion of the presentations in Session 3:

- To what degree did the development of fishing on the juvenile component of the stock from the 1950s to the 1970s influence the success of the catch of adults in the traps, in northern European waters, and the loss of age 5+ fish in the Bay of Biscay fishery?
- The utility of strong annual year classes to establish relationships between fishing grounds?
- It is important to know the contribution of recruits from the Mediterranean to the Atlantic fisheries:

  - Is there a wide variation in the proportion of fish leaving the Mediterranean?

- Does this proportion have any relationship with density-dependent effects (competition for space and for food)?
- As recovery rates in the Mediterranean are lower than in the east Atlantic, could this result in overestimating the number of Atlantic bluefin tuna leaving the Mediterranean?
- What are the general trends in distribution and movements of the juvenile component that leaves the Mediterranean?
- How important are learned behaviours and does the extirpation of abundances from other fishing grounds have strong implications in maintaining the presence of fish for the Bay of Biscay fisheries?

5.1.1.5 Session 4: Overall Vision of the Eastern Atlantic and Mediterranean Fisheries, Particularly the Traps

SCRS/2008/067. *Fromentin, J. M.* Back to the future: investigating historical data of bluefin tuna fisheries.
SCRS/2008/068. *Abid, N. and M. Idrissi*. Analysis of the Moroccan trap fishery targeting bluefin tuna (*Thunnus thynnus*) during the period 1986–2006.
SCRS/2008/069. *Bridges, C. R., O. Krohn, M. Deflorio and G. De Metrio*. Possible SST and NAO influences on the eastern bluefin tuna stock-the inexfish approach.
SCRS/2008/070. *Karakulak, F. S. and I. K. Oray*. Remarks on the fluctuations of bluefin tuna catches in Turkish waters.
SCRS/2008/071. *Vella, A.* Bluefin tuna (*Thunnus thynnus*) fisheries of the Maltese Islands in the central and southern Mediterranean Sea.
SCRS/2008/072. *Addis, P., I. Locci and A. Cau*. Anthropogenic impacts on the bluefin tuna (*Thunnus thynnus*) trap fishery of Sardinia (western Mediterranean).

Moderator-Speaker: J. M. Fromentin (IFREMER, France)

- In 1963 the leading bluefin tuna fisheries, which took place in the Norwegian Sea and North Sea, suddenly collapsed without any warning. While little is known of the reasons underlying this collapse, several hypotheses can be put forward, e.g. changes in bluefin tuna migratory routes, recruitment failure or eradication of a sub-population (all three hypotheses may be due to natural causes and/or overfishing).
- Current overexploitation in the Mediterranean Sea could explain why bluefin tuna did not return massively to the northeast Atlantic since the 1990s.
- An analysis of the Moroccan tuna trap fishery targeting bluefin tuna, which shows that the CPUE generally decreased from 1986 to 1995, increased during the period 1996 to 2001, and has since shown a downward trend.
- Historical data sets on catch and also model-generated data on spawning stock biomass (SSB) and recruitment have been used to look for possible influences of the North Atlantic oscillation (NAO) on the eastern bluefin tuna stock. Initial evidence has shown that total catch can be correlated to the winter NAO but only after a lag of two years.

- Turkish trap fisheries for bluefin date back to the 15th century. Fish traps used to be set in the Sea of Marmara, Bosphorus and in the Black Sea from April/May to late August. With the fall in fish stocks, marine pollution and urbanization, fish traps lost their importance in the Turkish bluefin tuna fishery. Recent studies show that bluefin tunas have not been migrating to and from the Black Sea since 1986.
- Traps in the Maltese Islands have caught bluefin tuna since 1748, reaching stable usage around 1948. However, this fishing method was finally replaced by longline.
- The traditional traps of Sardinia harvest the ancestral migratory flow of the Atlantic bluefin tuna at a fixed site. Therefore, it is reasonable to consider local perturbations generated by social and economic events and environmental changes as disruptive to the pathways of bluefin tuna schools and thus account for variability in the Mediterranean trap captures.

The Symposium concluded from Session 4:

- There was a strong connection between the Nordic fisheries and the northeast Atlantic traps (from Spain, Portugal and Morocco) and secondarily with Mediterranean traps as well as the northwest Atlantic trap. The collapse of the Nordic fisheries is not an isolated event.
- Atlantic bluefin tuna might be seen as a metapopulation made up of at least by three sub-populations that have varied in size in response to environmental changes and overfishing. Individual markers may be of great help to test the metapopulation hypothesis and thus the stock structure of bluefin tuna, as first results tend to show.
- Fishing grounds also changed significantly in the eastern Mediterranean Sea during the 20th century, moving from the Marmara Sea to the Black Sea and finally to the Aegean. In general, there were several extinctions/discoveries of important fishing grounds in the Mediterranean as well as in the east Atlantic during the 20th century.
- The importance of investigating fisheries on different spatial scales: i.e. large scales to detect connectivities between fisheries/stocks and impact of large-scale events (e.g. fishing, climate) and small-scale events to detect the impact of local events (e.g. coastal pollution due to industrial activities).
- Traps provide highly valuable scientific information from an ecological and a fisheries perspective as they are a passive fishing gear being set at the same location and submitted to low technical modifications.
- The causes of drastic changes in the fisheries of the 20th century are likely to result from interactions between biological, environmental, trophic and fishing processes.

5.1.1.6 Session 5: Overall Vision of the Western Atlantic Fisheries

SCRS/2008/073. *Takeuchi, Y., K. Oshima and Z. Suzuki*. Inference on the nature of Atlantic bluefin tuna off Brazil caught by the Japanese longline fishery around the early 1960s.

Moderator: J. Neilson (SABS, Canada)
Speaker: M. Lutcavage (UNH, U.S.A.)

- The western Atlantic fisheries also had a long historical record and provided some landings information from the late 1800s for the New England fishery.
- There is some evidence of a change in distribution to the northeast in the U.S. fishery. Possible reasons for the change could be related to prey distributions, and there have been observations of decreased condition of western bluefin tuna in the Gulf of Maine and southern Gulf of St. Lawrence.
- Japanese longline fishery which appeared suddenly and virtually disappeared in about 10 years with a substantial catch around the early 1960s. Among several hypotheses that were proposed to explain this event the authors favoured the temporal distribution hypothesis (similar to the concept of metapopulation).

The Session concluded by emphasizing that:

- The western stock is in a low state of abundance, and the spatial distribution may be changing.
- In common with the eastern stock, there are examples of tuna assemblages that have been extirpated. These have included large-scale aggregations, such as the one off Brazil in the 1960s, and smaller ones, such as the one that supported the trophy sport fishery off Nova Scotia (Sharpe Cup, 1930s–1960s).
- PSAT tagging results show diversity of movement patterns; areas where bluefin have been extirpated would be expected to be periodically revisited, raising the potential for recolonization.
- Changes in size structure can provide a warning of imminent fishery collapse.

5.1.1.7 General Discussion. Recommendations

Some important dynamics in Atlantic bluefin fisheries prior to 1970 should be incorporated into our overall analysis and utilized to shape our scientific advice to the Commission.

In the short-term (before the June 2008 stock assessment), it is unlikely that appropriate methodologies for incorporating historical information with different statistical characteristics into our stock assessment can be achieved to the full satisfaction of SCRS. This could only be achieved over a much longer period. While the current workplan for the 2008 bluefin stock assessment partly addresses the need to incorporate more biological realism into our evaluation of stock status, it is not clear that the current level of uncertainty in the assessment will be reduced in the short-run. Even in view of the high uncertainty, available information indicates that under recent conditions Atlantic bluefin tuna appear rapidly headed toward biological bankruptcy: spawning biomass is quickly shrinking and exploitation rates are much higher than the rate of interest nature provides. Our evaluations tell us that catches of bluefin tuna are now at their highest in ICCAT's history; biomass of age 8 and older bluefin tuna is at the lowest level ever estimated and possibly the lowest since

1950 or before, and that these catches are much too high to permit the Convention objectives to be achieved. More historical information is likely to better inform us as to the biomass levels to which rebuilding must occur to be consistent with the Convention objective and the rate at which rebuilding can occur to meet the Convention objective, but it is less likely to greatly alter our assessment of current exploitation rates. Critical to the issue is seeking an answer to the question: can bluefin tuna abundance levels previously observed off the coast of Brazil, the North Sea and traps be re-established?

In view of the above, it is incumbent upon SCRS to fully describe the information needed to progress in improving the advice we can offer the Commission and how to obtain such information. This is likely to involve coordinated data collection mechanisms which, in general, have not yet been realized through the various national programs underway.

An important aspect of recovering historical information which can better inform our assessment of bluefin is a data mining activity designed to capture and incorporate historical information into the ICCAT database. It is of key importance that SCRS has full access to all historical fishery data collected on bluefin tuna, especially those from the early years of the 20th century. This data mining should, for instance, target the recovery of all the historical data collected (published and unpublished) on the North Sea fisheries (e.g. within the ICES Tuna Working Group on Norwegian, Swedish, German fisheries and from all other potential sources), from the various traps active in the Atlantic and the Mediterranean Sea and the various bluefin fisheries that have been active during the period but not recorded in the ICCAT database. Data mining efforts should also target the recovery of the various sport fisheries that targeted bluefin tunas in the Atlantic and Mediterranean during the 20th century (allowing the identification of the place and dates of positive activities and the CPUE and sizes of fishes caught by each of these sport fisheries).

5.1.1.8 Adjournment

The General Director of the *IEO*, Dr. Enrique Tortosa, thanked the participants for their attendance and for the excellent work carried out. Dr. Tortosa pointed out the current false paradox between the exploitation of fishing resources and the conservation of the species. He recognized the importance of the conclusions of the Symposium and the SCRS' work to provide scientific advice. He further noted the relevance of the continued work of ICCAT in the adoption of measures for the management and conservation of the resources. The Director General of the *IEO* expressed the Spanish Government's wish that public policies in general be based on science and that economy and resources be considered jointly.

Dr. José Luis Cort, the General Coordinator of the Symposium, thanked the participants for the exceptional work carried out. He recognized that the Symposium had represented an excellent forum for discussion on subjects that are rarely discussed in other fora and he judged very positively the discussions held and the conclusions reached.

The Symposium was adjourned on April 24, 2008.

Fishing operations in the trap of Sancti-Petri (Spain), 1949

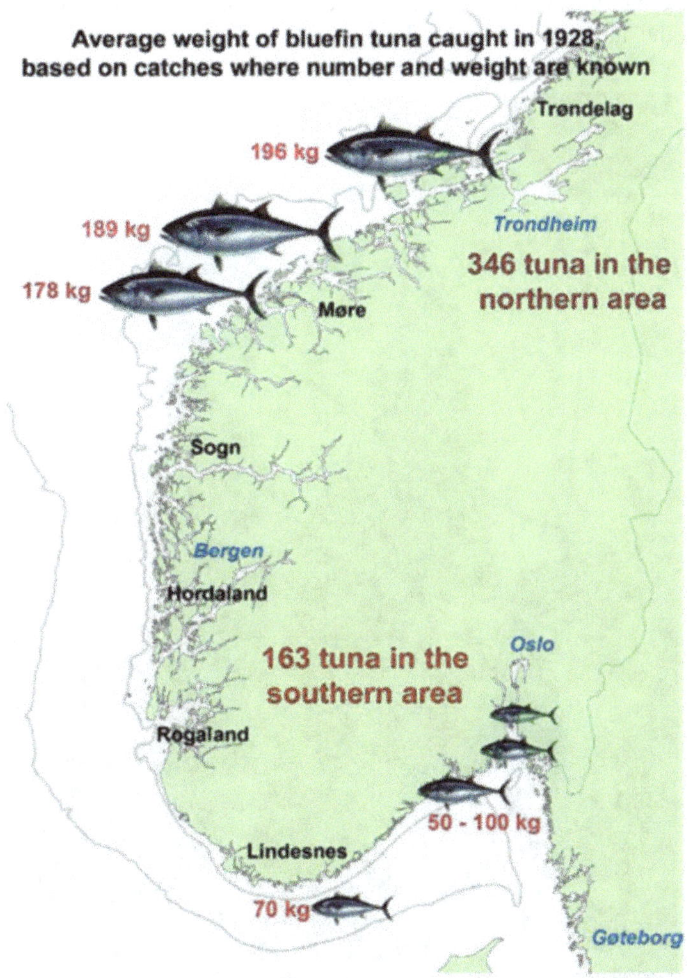

Fishing areas and mean weight of the tunas caught in Norway in 1928 (*Courtesy* of Øyvind and Magnus Tangen). Taken from ICCAT (2009); Tangen (2009)

Capture of bluefin tuna with purse seine in Norway (1928). Taken from ICCAT (2009); Tangen (1999, 2009), with the authorization of M. Tangen

Bluefin tuna fishing in the Bay of Biscay (1973) (http://www.photolib.noaa.gov/bigs/fish2077.jpg) (Documentary archive, *IEO*)

Bluefin tuna fishing in the traps (1982) (http://www.photolib.noaa.gov/bigs/fish2018.jpg) (Documentary archive, *IEO*)

Bluefin tuna, New Stocia (Canada), 1979 $W = 679$ kg, current Guinness world record Fraser (2008)

Cantabria

SANTANDER | 14 | La Ordenanza de Convivencia Ciudadana, que contempla fuertes multas, entró en vigor ayer

La crisis forzará a restringir más la captura mundial de atunes rojos

El presidente del ICCAT, Fabio Hazin, se mostró convencido de que la Comisión reducirá las capturas en su próxima reunión de noviembre

200 barcos españoles, de ellos 19 cántabros, se dedican a esta pesquería

Cien expertos de diversos países analizarán hasta el jueves el declive

Las conclusiones servirán de guía al comité científico y a la Comisión

Press note on the bluefin tuna symposium

Dr. Enrique Tortosa, Director General of the IEO, during the closing ceremony of the symposium (Documentary archive, *IEO*)

5.1.2 ICCAT-GBYP Project

According to its website (https://www.iccat.int/GBYP/en/) the project GBYP (Atlantic-wide research programme for bluefin tuna) was officially adopted by the ICCAT Commission in 2008. In 2009 the SCRS identified the priorities of the Research Plan as follows:

1. Improve basic data collection through data mining (including information from traps, observers, and VMS), developing methods to estimate sizes of fish caged, elaborating accurate CPUE indices for Mediterranean purse seine fleets, development of fisheries-independent information surveys and implementing a large scale scientific conventional tagging programme;
2. Improve understanding of key biological and ecological processes through electronic tagging experiments to determine habitat and migration routes, broad scale biological sampling of live fish and dead fish landed (e.g. gonads, liver, otoliths, spines, etc.), histological analyses to determine bluefin tuna reproductive state, biological and genetics analyses to investigate mixing and population structure; and ecological processes, including predator-prey relationships;
3. Improve assessment models and provision of scientific advice on stock status through improved modelling of key biological processes (including growth and stock-recruitment), further developing stock assessment models including mixing among areas, and developing and use of biologically realistic operating models for more rigorous management option testing.

During the Commission Meeting in 2009, a number of Contracting Parties expressed a willingness to make extra-budgetary contributions to such a programme with a view to initiating activities in 2009 related to different priorities: programme coordination, data mining, aerial surveys and tagging design studies; with additional research activities to be undertaken in the following years.

The provision to accept additional contributions from various entities and private institutions or companies was also agreed. In the same document, it was recommended to form a Steering Committee comprised by the SCRS Chair, the ICCAT Executive Secretary or his/her Assistant, bluefin tuna rapporteurs, and an outside expert with substantial experience in similar research undertakings for other tuna RFMOs, to guide and refine the Programme as necessary.

GBYP (Grand Bluefin Tuna Year Programme) was adopted as the official acronym of the research, which was initiated at the end of March 2010. It was divided in annual phases, sometimes covering parts of following years.

The project ICCAT-GBYP has meant and continues to mean the implementation of numerous scientific activities that contribute enormously to the knowledge of ABFT. These activities can be consulted under the tab 'Research' of the project web page.

Dr. Antonio Di Natale was the project coordinator between 2010 and 2018. Dr. Francisco Alemany took over from him in February 2018.

5.1.3 *Symposium on Bluefin Tuna Traps Fisheries (ICCAT-GBYP 2011)*

The following is a literal transcription of some of the contents of the report (ICCAT 2012).

The aim of the Symposium was to discuss and review the information from tuna traps to maximize the use of scientific information provided by this traditional gear. The Symposium should also provide the opportunity to improve knowledge and achieve a better understanding of this ancient fishery in order to find a common methodological approach to standardize the CPUEs related to this fishing activity, which has the longest historical series among all the world's fisheries. Finding a common standardization procedure will allow these data series to be better used in the stock assessment models. Thus, this Symposium also represents an excellent opportunity to define the potential contribution of the trap fishery to the future work of the SCRS on the bluefin tuna stocks which should further help to improve the current management and conservation measures for bluefin tuna fisheries.

The Symposium was a response to a recommendation made by the Standing Committee on Research and Statistics (SCRS) in 2010 and was carried out within the framework of the ICCAT Atlantic-wide Research Program on Bluefin Tuna (GBYP).

The Symposium was open on May 23, 2011 by Dr. Pilar Pallarés, ICCAT Assistant Executive Secretary. Dr. Pallarés, on behalf of the Executive Secretary of ICCAT, thanked the Government of Morocco, the authorities of 4 Tangier and the National Institute of Marine Research for hosting the meeting, and welcomed the participants.

Dr. Pallarés made reference to the ICCAT Atlantic-wide Research Program on Bluefin Tuna (GBYP) within whose framework the Symposium was carried out and emphasized the importance of the trap fishery as a source of information on the history of bluefin tuna and its contribution to better understand the long-term dynamics of the population of this species. Dr. Pallarés expressed the wish that the Symposium will help the SCRS better assess the bluefin tuna stock and therefore contribute towards improving the management of the stocks.

Dr. José Luis Cort, Chair of the Symposium Scientific Committee, acted as Moderator for the Symposium. Dr. Cort acknowledged the kind support provided by the Moroccan tuna traps and the tuna industry for the organization of the Symposium and for hosting the official dinner. The Symposium was organized into four thematic sessions, each coordinated by a Moderator. The following participants acted as Moderators and Rapporteurs:

Moderators and Rapporteurs:

Session 1: A. Di Natale, M. Idrissi (ICCAT-GBYP)
Session 2: M. Ortiz (ICCAT)
Session 3: A. Fonteneau (IRD, France)
Session 4: Z. Suzuki (AFFRC, Japan).

A total of 58 scientists attended the Symposium.
The documents of the symposium are available at ICCAT (2012).

5.1.3.1 Session 1: Historical, Cultural and Technological Aspects of Bluefin Tuna Trap Fisheries

The first Session included a total of eight presentations covering various aspects of this essential background.

SCRS/2011/036. *Di Natale, A.* An iconography of tuna traps: Essential information for the understanding of the technological evolution of this ancient fishery.
SCRS/2011/078. *Garcia Garcia, F.* Las almadrabas de la costa andaluza bajo el dominio de la casa ducal de Medina Sidonia. Su tipología, sus producciones y sus problemáticas.
SCRS/2011/071. *Gil Pereira, J.* Historical bluefin tuna catches from southern Portugal traps.
SCRS/2011/083. *Manfrin, G., Mangano, A., Piccinetti, C., Piccinetti, R.* Les données sur la capture des thons par les madragues dans l'archive du Prof. M. Sella.
SCRS/2011/069. *Farrugio, H.* Données historiques sur les anciennes madragues françaises de Méditerranée.
SCRS/2011/081. *Abid, N. Benchoucha, S., Belcaid, S., Lamtai, A. and El Fanichi, C.* Moroccan tuna traps: History and current situation.
SCRS/2011/031. *López González, J.A., Ruiz Acevedo, J.M.* Series históricas de capturas del bluefin tuna en las almadrabas del Golfo de Cádiz (Siglos XVI-XXI).
SCRS/2011/037. *Di Natale, A.* Literature on the eastern Atlantic and Mediterranean tuna trap fishery.

The discussion examined the fact that not all the Mediterranean areas are documented in the same way and the reason for that possibly reflects the fact that the trap fishery was perhaps not practiced in some east Mediterranean countries. Some data were provided in the 13th century by the famous geographer El Idrissi. It was noted that the detailed archives of the Ottoman Empire have not yet been analysed. The discussion also included historical information on the impact of killer whales on the fishery in the Strait of Gibraltar, the socioeconomic importance of the direct and indirect income derived from the tuna trap industry and the effects of having a series of tuna traps along the same coast. The high catches in the historical series are understandable, but it is more difficult to understand the reasons for the low catches. The mixture of species that are reported in the historical Duke of Medina Sidonia series will require additional data mining for further study. The conclusion of the session stressed the relevant importance of tuna traps from a cultural, historical and social point of view and the need to preserve this activity for the future. It was noted that historical data on trap fisheries require the identification of size classes, common names, size-class category, possible different species composition and market categories recorded, before combining multiple sets of data for these analyses.

5.1.3.2 Session 2: Bluefin Tuna Trap Data and Standardization for Assessment Uses

The use of bluefin tuna trap fisheries data on catch, fishing effort and auxiliary information for assessment and stock assessment was presented and discussed during this session. A total of five papers were presented:

SCRS/2011/039. *Ortiz, M., Palma, C., Pallarés, P., Kell, L., Idrissi, M. and Di Natale, A.* Tuna trap data existing in the ICCAT database and data recovered under the GBYP.
SCRS/2011/028. *Cort, J.L., de la Serna, J. M. and Velasco, M.* Annual mean weight of bluefin tuna (*Thunnus thynnus*) caught by the traps in the south of Spain between 1914–2010.
SCRS/2011/038. *Di Natale, A. and Idrissi, M.* Factors to be taken into account for a correct reading of tuna trap catch series.
SCRS/2011/027. *Abid, N., de la Serna, J. M., Rodriguez Marín, E., Macías, D., Rioja, P., Ortiz de Urbina, J.M.* Standardized CPUE of bluefin tuna (*Thunnus thynnus*) caught by Moroccan and Spanish traps for the period 1981–2009.
SCRS/2011/041. *Kai, M. and Takeuchi, Y.* Abundance index of young Pacific bluefin tuna (*Thunnus orientalis*) estimated from the Japanese set net fishery's data.

The general discussion commented on the importance of the trap fisheries as a major source of scientific and fisheries data for the assessment of bluefin tuna stocks and the need to continue supporting the scientific collection and reporting of data from trap operations. It was noted that management measures, such as quota restrictions, need to be taken into account for data collection and data analysis. Therefore, there is a need to record retained as well as released fish from traps, and to have standard descriptions of fishing operations to define clearly the fishing effort unit. In addition to the CPUE series from trap catches, size and age distribution samples should be routinely collected. Analyses from traps in southern Spain have shown the change in the trends of size and age of fish caught among various series of years. In such cases it is recommended that the CPUE series be standardized by age or age-groups for the assessment process. It was also recommended that analyses of trap fishery data be carried out on a wider regional scale, including coordination of the potential links between overall environmental trends (currents, decadal oscillations, etc.).

It is also important to understand the effects of small scale meteorological events on the catch rates of local traps.

5.1.3.3 Session 3: Tuna Traps as a Reliable Scientific Observatory for Bluefin Tuna Stocks

Six presentations were given during this session. They presented results of scientific value obtained from a wide range of historical and currently active traps, mainly from the Mediterranean Sea.

SCRS/2011/075. *Addis, P., Secci, M., Locci, I., Sabatini, A., Dean, J. M. and Cau, A.* Long-term analysis (1993–2010) of the catches of Atlantic bluefin tuna (*Thunnus thynnus*) from the traditional trap fisheries of Sardinia.
SCRS/2011/0074. *Dean, J. M., Andrushchenko, I. and Neilson, J. D.*The western Atlantic bluefin tuna trapnet fishery.
SCRS/2011/042. *Suzuki, Z. and Kai, M.* Movement of Atlantic bluefin tuna toward the Strait of Gibraltar inferred from Japanese longline data.
SCRS/2011/029. *De la Serna, J. M., Macías, D., Ortiz de Urbina, J. M., Rodríguez-Marín, E., Abascal, F.* Study on the eastern Atlantic and Mediterranean bluefin tuna stock using the Spanish traps as scientific observatories.
SCRS/2011/084. *Fonteneau, A.* Mediterranean traps in the 21st century: Research tools for the conservation of bluefin tuna.
SCRS/2011/080. *Cannas, R., Ferrara, G., Milano, I., Landi, M., Cariani, A., Addis, P., Cau, A., Piccinetti, C., Sella, M., Tinti, F.* Spatio-temporal genetic variation of Atlantic bluefin tunas from Sardinian and Mediterranean tuna traps.

The general conclusion following these presentations and the discussion was a firm consensus that for ICCAT scientists bluefin tuna traps were for four centuries, and remain today, an invaluable gold-mine of data (statistical, biological and others), while recent fishing mortality due to tuna traps is low and the sizes of bluefin tuna caught by the traps is close to the optimum in terms of yield per recruit. On the other hand, ICCAT scientists recognize that the statistical and scientific information obtained from modern and recent gears (such as purse seiners) is and has been extremely limited, particularly for a period during which these modern gears have been the major source of excessive fishing mortality suffered by the bluefin tuna stock.

In this context, these data from traps are as essential today as they were in the past for ICCAT bluefin tuna stock assessment, because they provide age-specific measures of stock biomass for both the sedentary and migrating fraction of the bluefin tuna stock, and also a wide range of biological data that constitute a very important component in the bluefin tuna stock assessment models.

There is a consensus among scientists that it would be very negative to stop now, after 400 years of continuous data, these unique statistical series from the trap fisheries. The recommendation from this Symposium to the ICCAT Commission and to the ICCAT CPCs is to maintain the trap fisheries operational, *inter alia* because of their high value for scientific research and stock assessment. The Symposium participants also recommended that these traps should be kept open for a period that allows the continuation of their long-term statistical series.

It is also recommended that these traps be considered as "ICCAT tuna observatories", increasing their cooperation with ICCAT and its scientific programs by providing full access to their detailed catch and effort data, giving access to biological sampling and allowing the tag and release of bluefin tunas.

5.1.3.4 Session 4: Tuna Traps and Bluefin Tuna: Socio-Economy, Global Management and Market Issues

There were four documents presented at this Session:

SCRS/2011/040. *Suzuki, Z., Kai, M.* General information on Japanese trap fisheries catching Pacific bluefin tuna (*Thunnus orientalis*): Fishery and socio-economic roles.
SCRS/2011/082. *Malouli Idrissi, M., Zahraoui, M. Nhhala, H.* Les madragues au Maroc: Aspects économiques.
SCRS/2011/076. *Addis, P., Secci, M., Locci, I., Cannas, R., Greco, G., Dean, J. M., Cau, A.* Social, cultural and basic economic analysis of the trap fishery of Sardinia: First step towards parameterization.
SCRS/2011/077. *Addiss, P., Secci, M., Locci, I. Cau, A.* Harvesting, handling practices and processing of bluefin tuna captured in the trap fishery: Possible effects on the flesh quality.

Discussions centred on the following:

As concerns the Japanese trap presentation, it was asked whether or not the Japanese government gave preferential treatment to traditional fisheries such as the traps with respect to management of the stock. The answer was that it did not give any special treatment to the traditional fisheries but there appeared to be general agreement that the traditional fisheries should be less affected by the regulatory measures because their fisheries had been sustainably conducted.

Clarification was made to the fact that fuel cost is a dominant item of expenditure in the Moroccan trap fishery and it was explained that the reason for the high fuel cost in total expenditure was due to the long distance between the ports and trap locations and to the trap setting-up operations, which require considerable time and effort.

Some concern was raised about using the "matanza" (killing of bluefin in the final stage of trap harvest) as a tourist attraction. Although there were no complaints from the tourists who watched the "matanza", it may be necessary for persons in charge of the trap fishery to be prepared to give an explanation of this activity.

A question was raised about the cooperation among trap owners and scientists who collect biological samples and the response was that most of the sampling is permitted although it depends to a large extent on the kind of samples taken. It was noted that maintaining a good relationship with the industry is important and this may be done through feedback to the industry from the scientific research results obtained by biological sampling. It was agreed that the proposals made by the tuna trap industry be included as Appendix 3 of the report in a summarized format.

5.1.3.5 Recommendations

- The historical data series from the tuna trap fishery archives that have been recovered in the last two years provide an important improvement to the ICCAT data base. The Symposium recommends that further details be made available by national scientists for a better understanding of the natural fluctuations of the stock and to improve standardised CPUEs taking into account the most relevant variables.

- The considerable historical and cultural importance of the tuna trap fishery and industry shall be preserved. The Symposium recommends that the national governments concerned take the necessary steps to promote the urgent conservation of the few remaining tuna traps by considering, among others, the possibility of requesting their inclusion under "World Cultural Heritage" by UNESCO.
- It is also recommended that these traps be considered as "ICCAT Tuna Observatories" by increasing their full cooperation with ICCAT and its scientific programs, providing full access to their detailed catch and effort data, permitting biological sampling to be carried out, and by allowing the tag and release of bluefin tunas.
- For the opportunity to effectively use the tuna traps as "Tuna Scientific Observatories" the Symposium reiterates the recommendation from the SCRS in 2010 to the Commission to establish a scientific quota allocation for the ICCAT Atlantic-Wide Research Program on Bluefin Tuna (GBYP). This allocation should not fall under the restrictions of current size regulations and should include all fish-size ranges.
- For standardizing the CPUE series from trap fisheries, it is recommended that:
  - Records be kept of landed fish as well as fish released from the traps.
  - Records be kept of size and/or age information of the fish caught, and indices be developed by age or age groups if there are changes in the size distribution of fish caught in the traps.
  - Regional-wide studies be promoted on the trends of catch rates at size-age from different tuna traps.
- There is a consensus among scientists that it would be very negative to stop now, following 400 years of continuous data, these unique statistical series from the trap fisheries. The recommendation from this Symposium to the Commission and to the ICCAT CPCs is to maintain the trap fisheries operational, *inter alia,* because of their high value to scientific research and stock assessment.
- The Symposium participants also recommended that these traps be kept open for a time period long enough to maintain the consistency of their long-term statistical series.

5.1.3.6 Adoption of the Report and Closure

Dr. José Luis Cort, the Moderator of the Symposium, thanked the participants for the exceptional work carried out and expressed special appreciation to the Secretariat and, in particular, to the GBYP for the excellent organization of the Symposium.

The report of the Symposium was adopted. The Symposium was adjourned on May 25, 2011.

Following this general presentation of ABFT and its fishing, the aim of the present article is to study the Bay of Biscay fishery in detail. This fishery is one of those that has been the subject of the largest number of studies and the authors of the present paper have recently published a study on it (Cort and Abaunza, 2015) describing the interaction of the fishery with the other spawner fisheries in the eastern Atlantic.

Representation of bluefin tuna fishing in Cádiz (16th century). Taken from Di Natale (2012); ICCAT (2012)

End phase in the trap fishing operation: *la levantada* ("the lifting"), 1982 (Documentary archive, *IEO*)

“Death chamber” (2009)

Raising the tunas with hooks (1982) (http://www.photolib.noaa.gov/bigs/fish2017.jpg) (Documentary archive, *IEO*)

Tuna trap of Barbate, 2009 (Cádiz, Spain) (Documentary archive, *IEO*)

Bluefin tuna catch in the trap of Barbate, 2009 (Cádiz, Spain) (Documentary archive, *IEO*)

Landing the catches of the trap of Barbate, 2009 (Cádiz, Spain) (Documentary archive, *IEO*)

References

Abascal F, Medina A, de la Serna JM, Godoy D, Aranda G (2016) Tracking bluefin tuna reproductive emigration into the Mediterranean Sea with electronic pop-up satellite archival tags using two tagging procedures. Fish Oceanogr 25(1):54–66

Addis P, Secci M, Biancacci C, Loddo D, Cuccu D, Palmas F, Sabatini A (2016) Reproductive status of Atlantic bluefin tuna, *Thunnus thynnus*, during migration off the coast of Sardinia (western Mediterranean). Fish Res 181:137–147

Ailloud LE, Lauretta MV, Hanke A, Golet W, Allman R, Siskey MR, Secor DH, Hoening JM (2017) Improving growth estimates for Western Atlantic bluefin tuna using an integrated modeling approach. Fish Res 191:17–24

Alemany F, Quintanilla L, Vélez-Belchí P, García A, Cortés D, Rodríguez JM et al (2010) Characterization of the spawning habitat of Atlantic bluefin tuna and related species Balearic Sea (western Mediterranean). Prog. Oceanogr 86:21–38

Álvarez-Berastegui D, Ciannelli L, Aparicio-Gonzalez A, Reglero P, Hidalgo M et al (2014) Spatial scale, means and gradients of hydrographic variables define pelagic seascapes of bluefin and bullet tuna spawning distribution. PLoS ONE 9 (10):e109338. https://doi.org/10.1371/journal.pone.0109338[pmc free article][PubMed]

Anonymous (1994) An assessment of Atlantic bluefin tuna. National Academy Press, Washington, D.C. 148 p

Aranda G, Abascal FJ, Varela JL, Medina A (2013) Spawning behaviour and post-spawning migration patterns of Atlantic bluefin tuna (*Thunnus thynnus*) ascertained from satellite archival tags. PLoS ONE 8(10):e76445. https://doi.org/10.1371/journal.pone.0076445

Baglin REJ (1982) Reproductive biology of western Atlantic bluefin tuna. Fish Bull 80:121–134

Battaglia P, Andaloro F, Consoli P, Esposito V, Malara D, Musolino S, Pedà C (2013) Feeding habits of the Atlantic bluefin tuna, *Thunnus thynnus* (L. 1758), in the central Mediterranean Sea (Strait of Messina). Helgol Mar Res 67(1):97–107. https://doi.org/10.1007/s10152-012-0307-2

Block BA, Teo SLH, Walli A, Boustany A, Stikesbury MJW, Farwell CJ, Weng KC, Dewar H, Williams TD (2005) Electronic tagging and population structure of Atlantic bluefin tuna. Nature 434:1121–1127

Brêthes JC (1978) Campagne de marquage de jeunes thons rouges au large des côtes du Maroc. Col Vol Sci Pap ICCAT 7:313–317

Brêthes JC (1979) Sur les premiers recuperations des thons rouges marqués en juillet 1977 au large du Maroc. Col Vol Sci Pap ICCAT 8:367–369

Brêthes JC, Mason JM Jr (1979) Bluefin tuna tagging off the Atlantic coast of Morocco in 1978. Col Vol Sci Pap ICCAT 8:329–332

Brodziak J, Ianelli J, Lorenzen K, Methot RD Jr (2011) Estimating natural mortality in stock assessment applications. NOAA Technical Memorandum NMFS-F/SPO-119; 48 p

Brophy D, Haynes P, Arrizabalaga H, Fraile I, Fromentin JM, Garibaldi F, Katavic I, Tinti F, Karakulak S, Macías D, Busawon D, Hanke A, Kimoto A, Sakai O, Deguara S, Abid N, Neves Santos M (2015) Otolith shape variation provides a marker of stock origin for north Atlantic bluefin tuna (*Thunnus thynnus*). Mar Freshw Res. http://dx.doi.org/10.1071/MF15086

Butler MJA, Accy JF, Dickson CA, Hunt JJ, Burnet CD (1977) Apparent age and growth, based on otoliths analysis, of giant bluefin tuna (*Thunnus thynnus thynnus*) in the 1975–76 Canadian catch. Col Vol Sci Pap ICCAT 6:318–330

Butler CM, Logan JM, Pravaznik JM, Hoffmayer ER, Staudinger MD, Quattro JM, Roberts MA, Ingram GW Jr, Pollack AG, Lutcavage ME (2015) Atlantic bluefin tuna *Thunnus thynnus* feeding ecology in the northern Gulf of Mexico: a preliminary description of diet from the western Atlantic spawning grounds. J Fish Biol 86(1):365–374. https://doi.org/10.1111/jfb.12556

Caddy JF, Dickson CA, Butler JA (1976) Age and growth of giant bluefin tuna (*Thunnus thynnus thynnus*) taken in Canadian waters in 1975. J Fish Res Board Can, MS Rep. No. 1395

Cermeño P, Quílez-Badia G, Ospina-Álvarez A, Sainz-Trápaga S, Bustany AM, Seitz AC, Tudela S, Block BA (2015) Electronic tagging of Atlantic bluefin tuna (*Thunnus thynnus*, L.) reveals hábitat use and behaviors in the Mediterranean Sea. PLoS ONE 10(2):e0116638. https://doi.org/10.1371/journal.pone.0116638

Compeán-Jiménez G, Bard FX (1983) Growth increments on dorsal spines of eastern Atlantic bluefin tuna (*Thunnus thynnus* (L.)) and their possible relation to migrations patterns. NOAA, Tech Rep NMFS 8:77–86

Corriero A, Kakakulak S, Santamaría N, Deflorio M, Spedicato D, Addis P, Fenech-Farrugia A, Vassallo-Agius R, de la Serna JM, Oray I, Cau A, De Metrio G (2005) Size and age at sexual maturity of femelle bluefin tuna (*Thunnus thynnus* L. 1758) from the Mediterranean Sea. J Appl Ichthyol 21:483–486

Cort JL (1990) Biología y pesca del atún rojo, *Thunnus thynnus* (L.), del mar Cantábrico. Doctoral thesis. Publicaciones especiales, IEO, vol 4, 272 p

Cort JL, Arregui I, Estruch V, Deguara S (2014) Validation of the growth equation applicable to the eastern Atlantic bluefin tuna, *Thunnus thynnus* (L.), using $L_{\max}$, tag-recapture and first dorsal spine analysis. Rev Fish Sci Aquac 22(3):239–255. https://doi.org/10.1080/23308249.2014.931173

Cort JL, Deguara S, Galaz T, Mèlich B, Artetxe I, Arregi I et al (2013) Determination of $L_{\max}$ for Atlantic bluefin tuna, *Thunnus thynnus* (L.), from meta-analysis of published and available biometric data. Rev Fish Sci 21(2):181–212. https://doi.org/10.1080/10641262.2013.793284

Cort JL, Estruch VD, Santos MN, Di Natale A, Abid N, de la Serna JM (2015) On the variability of the length-weight relationship for Atlantic bluefin tuna, *Thunnus thynnus* (L.). Rev Fish Sci Aquac 23(1):23–38. https://doi.org/10.1080/23308249.2015.1008625

Cort JL, Nøttestad L (2007) Fisheries of bluefin tuna (*Thunnus thynnus*) spawners in the Northeast Atlantic. Col Vol Sci Pap ICCAT 60:1328–1344

De Buen F (1925) Biología del atún, *Orcinus thynnus* (L.). Resultado de las campañas realizadas por acuerdos internacionales, 1. Madrid, 118 p

De la Serna JM, Abid AN, Godoy D (2013) Posible influencia sobre el comportamiento migratorio del atún rojo (*Thunnus thynnus*) de las distintas estrategias de marcado electrónico utilizadas en las almadrabas y jaulas de engorde. Col Doc Sci Pap ICCAT, 69:427–434

De Metrio G, Arnold GP, Block BA, de la Serna JM, Deflorio M, Cataldo M, Yannopoulos C, Megalofonou P, Beeper S, Farwell C, Seitz A (2002) Behaviour of post-spawning Atlantic bluefin tuna tagged with pop-up satellite tags in the Mediterranean and eastern Atlantic. Col Vol Sci Pap ICCAT 54(2):415–424

Dicenta A, Piccinetti C (1980) Comparison between the estimated reproductive stocks of bluefin tuna (*T. thynnus*) of the Gulf of Mexico and western Mediterranean. Col Vol Sci Pap ICCAT 9(2):442–448

Di Natale A (2012) Literature on eastern Atlantic and Mediterranean tuna trap fishery. Col Vol Sci Pap ICCAT 67:175–220

Di Natale A, Tensec S, Pagá A (2016) Preliminary information about the ICCAT GBYP tagging activities in phase 5. Col Vol Sci Pap ICCAT 72(6):1589–1613

Druon JN, Fromentin JM, Aulanier F, Heikkonen J (2011) Potential feeding and spawning habitats of Atlantic bluefin tuna in the Mediterranean Sea. Mar Ecol Prog Ser 439:223–240. https://doi.org/10.3354/meps09321

Druon JN, Fromentin JM, Hanke AR, Arrizabalaga H, Damalas D, Tičina V et al (2016) Habitat suitability of the Atlantic bluefin tuna by size class: an ecological niche approach. Progr Oceanogr 142:30–46. https://doi.org/10.1016/j.pocean.2016.01.002

Duclerc J, Sachi J, Piccinetti-Manfrin G, Piccinetti C, Dicenta A, Barrois JM (1973) Nouvelles données sur la reproduction du thon rouge (*Thunnus thynnus*) et d´autres espèces des thonidés en Mediterranée. Rev Trav Inst Pêches Marit 37(2):163–176

Estrada JA, Lutcavage M, Thorrold SR (2005) Diet and trophic position of Atlantic bluefin tuna (*Thunnus thynnus*) inferred from stable carbon and nitrogen isotope analysis. Mar Biol 147:37–45. https://doi.org/10.1007/s00227-004-1541-1

Florido del Corral D (2013) Las almadrabas andaluzas bajo el consorcio nacional almadrabero (1928–1971): aspectos socio-culturales y políticos. Semana, Ciencias Sociais e Humanidades 25:117–151. ISSN 1137-9669

Florido del Corral D, Santos A, Ruiz JM, López JA (2018) Las almadrabas suratlánticas andaluzas. Historia, tradición y patrimonio (siglos XVIII–XXI). Editorial Universidad se Sevilla, 328 p. ISBN 978-84-472-1885-1

Fonteneau A, Le Person A (2009) Bluefin fishing in Lannion Bay, northern Brittany, during the 1946–1953 period. Col Vol Sci Pap ICCAT 63:69–78

Fonteneau A, Maguire JJ (2014) On the natural mortality of eastern and western bluefin tuna. Col Vol Sci Pap ICCAT 70(1):289–298

Frade F (1938) Recherches sur la maturité sexuelle du thon rouge de l´Atlantique et de la Mediterranée. Bul Soc Portug Sci Nat 12:243–250

Fraile I, Arrizabalaga H, Rooker R (2014) Origin of Atlantic bluefin tuna (*Thunnus thynnus*) in the Bay of Biscay. ICES J Mar Sci. https://doi.org/10.1093/icesjms/fsu156

Fraser K (2008) *Possessed. World Record Holder for Bluefin Tuna.* Kingstown, Nova Scotia: T & S Office Essentials and printing, 243 p

Fromentin JM (2002) Final report of STROMBOLI-EU-DG XIV project 99/022. In: European Community-DG XIV, Brussels, 109 p

Fromentin JM (2009) Lessons from the past: investigating historical data from bluefin tuna fisheries. Fish Fish 10:197–216

Fromentin JM, Bonhommeau S, Arrizabalaga H, Kell LT (2014) The spectre of uncertainty in management of exploited fish stocks: the illustrative case of Atlantic bluefin tuna. Marine Policy 47:8–14

Fromentin JM, Powers J (2005) Atlantic bluefin tuna: population dynamics, ecology, fisheries and management. Fish Fish 6:281–306

Fromentin JM, Restrepo V (2009) A year-class curve analysis to estimate mortality of Atlantic bluefin tuna caught by the Norwegian fishery from 1956–1979. Col Vol Sci Pap ICCAT 64(2):480–490

Galuardi B, Logan JM, Neilson JD, Lutcavage M (2015) Complex migration routes of Atlantic bluefin tuna (*Thunnus thynnus*) question current population structure paradigm. Can J Fish Aquatic Sci 966–976

García A, Alemany F, de la Serna JM, Oray I, Karakulak S, Rollandi L, Arigo A, Mazzola S (2005) Preliminary results of the 2004 bluefin tuna larval surveys off different Mediterranean sites (Balearic Archipelago, Levantine Sea, and the Sicilian Channel). Col Vol Sci Pap ICCAT 58:1420–1428

García A, Cortés D, Ramírez T, Fehri-Bedoui R, Alemany F, Rodríguez JM, Carpena Á, Álvarez JP (2006) First data on growth and nucleic acid and protein content of field-captured Mediterranean bluefin (*Thunnus thynnus*) and albacore (*Thunnus alalunga*) tuna larvae: a comparative study. Scientia Marina 70(S2). https://doi.org/10.3989/scimar.2006.70s267

Goñi N, Fraile I, Arregui I, Santiago J, Boyra G, Irigoien X, Lutcavage M et al (2010) On-going bluefin tuna research in the Bay of Biscay (Northeast Atlantic): The "Hegalabur 2009" project. Col Vol Sci Pap ICCAT 65(3):755–769

Goñi N, Onandia I, López J, Arregui I, Uranga J, Melvin GD, Boyra G, Arrizabalaga H, Santiago J (2017) Acoustic-based fishery-independent abundance index of juvenile bluefin tunas in the Bay of Biscay. Col Vol Sci Pap ICCAT 73(6):2044–2057

Heinisch G, Rosenfeld H, Knapp JM, Gordin H, Lutcavage ME (2014) Sexual maturity in western Atlantic bluefin tuna. Sci Rep 4, Article number: 7205. https://doi.org/10.1038/srep07205

ICCAT (1996) National report of Japan. ICCAT Report for biennial period, 1996–97. Part I, vol. 2, 204 p. http://iccat.int/Documents/BienRep/REP_EN_96-97_I_2.pdf

ICCAT (2007) Report for biennial period, 2006–07. Part I (2006). Vol. 2, 240 p. https://www.iccat.int/Documents/BienRep/REP_EN_06-07_I_2.pdf

ICCAT (2008) Report for biennial period, 2006–07. Part II (2007). Vol. 1, 276 p

ICCAT (2009) Report of the world symposium for the study into the stock fluctuation of northern bluefin tunas (*Thunnus thynnus* and *Thunnus orientalis*), including the historical periods. Col Vol Sci Pap ICCAT 63:1–49

ICCAT (2010) ICCAT Manual. Description of species. Chapter 2; 2.1.5 Atlantic bluefin tuna, vol 99. Madrid, ICCAT, pp 93–111 http://iccat.int/Documents/SCRS/Manual/CH2/2_1_5_BFT_ENG.pdf

ICCAT (2012) ICCAT-GBYP symposium on trap fisheries for bluefin tuna. Col Vol Sci Pap ICCAT 67:3–30

ICCAT (2017) Report of the 2017 ICCAT bluefin stock assessment meeting. Madrid, Spain, 22–27 July 2017, 106 p. http://iccat.int/Documents/Meetings/Docs/2017_BFT_ASS_REP_ENG.pdf

Hamre J (1960) Tuna investigation in Norwegian coastal waters 1954–1958. Ann Biol Cons Int Expl Mer 15:197–211

Hamre J (1965) The Norwegian tuna investigations, 1962–1963. Ann Biol Cons Int Expl Mer 20:232–236

Hamre J, Tiews K (1963) Second report of the bluefin tuna working group. ICES, C.M. 1963 Scombriform Fish Committee, no. 14, 29 p

Karakulak S, Oray I, Corriero A, Deflorio M, Santamaria N, Desantis S, De Metrio G (2004) Evidence of a spawning area for the bluefin tuna (*Thunnus thynnus*) in the Eastern Mediterranean. J Appl Ichthyol 20:318–320

Laiz-Carrión R, Gerard T, Uriarte A, Malca E, Quintanilla JM, Muhling B, Alemany F, Privoznik SL, Shiroza A, Lamkin JT, García A (2015) Trophic ecology of Atlantic bluefin tuna (*Thunnus thynnus*) larvae from the Gulf of Mexico and NW Mediterranean spawning grounds: a comparative stable isotope study. PLoS ONE 10(7):2015. https://doi.org/10.1371/journal.pone0133406

Lamboeuf M (1975) Contribution a la connaissance des migrations des jeunes thons rouges a partir du Maroc. Col Vol Sci Pap ICCAT 4:141–144

Logan JM, Rodríguez-Marín E, Goñi N, Barreiro S, Arrizabalaga H, Golet WJ, Lutcavage M (2010) Diet of young Atlantic bluefin tuna (*Thunnus thynnus*) in eastern and western foraging gound. Mar Biol 12. https://doi.org/10.1007/s00227-010-1543-0

Lozano F (1958) Los escómbridos de las aguas españolas y marroquíes y su pesca. Trab Inst Esp Ocean 25:254 p

Lozano F (1959) The usu of echo-sondeurs in the study of the migrations of tuna. Proc Gen Fish Coun Medit 5:101–104

Luque P, Rodríguez-Marín E, Landa J, Ruiz M, Quelle P, Macías D, Ortiz de Urbina JM (2014) Direct ageing of *Thunnus thynnus* from the eastern Atlantic Ocean and western Mediterranean Sea using dorsal fin spines. J Fish Biol 84:1876–1903

Lutcavage M, Kraus SD (1995) The feasibility of direct photographic aerial assessment of giant bluefin tuna in New England waters. Fish Bull 93:495–503

Lutcavage M, Kraus SD, Hoggard W (1997) Aerial assessment of giant bluefin tuna in the Bahama Banks-Straits of Florida, 1995. Fish Bull 95:300–310

Lutcavage M, Galuardi B, Lam TCH (2013) Predicting potential Atlantic spawning grounds of western Atlantic bluefin tuna based on electronic tagging results, 2002–2011. Col Vol Sci Pap ICCAT 69:955–961

MacKenzie BR, Payne MR, Boje J, Hoyer JL, Siegstad H (2014) A cascade of warming impacts brings bluefin tuna to Greenland waters. Glob Change Biol 20(8):2484–2491

Manfrin G, Mangano A, Piccinetti C, Piccinetti R (2012) Les données sur la capture des thons par les madragues dans l´archive du prof Sella. Col Vol Sci Pap ICCAT 67:106–111

Mather III FJ, Mason Jr JM, Jones AC (1973) Distribution fisheries and life history data relevant to identification of Atlantic bluefin tuna stocks. Col Vol Sci Pap ICCAT 2:234–258

Mather III FJ, Mason Jr JM, Jones AC (1995) Historical document: life history and fisheries of Atlantic bluefin tuna. In: NOAA technical memorandum, NMFS-SEFSC-370, Miami Fl, 165 p

Medina A, Abascal FJ, Megina C, García A (2002) Stereological assessment of the reproductive status of female Atlantic northern bluefin tuna during migrations to Mediterranean spawning grounds through the Strait of Gibraltar. J Fish Biol 60:203–217. https://doi.org/10.1111/j.1095-8649.2002.tb02398.x

Medina A, Cort JL, Aranda G, Varela JL, Aragón L, Abascal F (2011) Summary of bluefin tuna tagging activities carried out between 2009 and 2010 in the East Atlantic and Mediterranean. Col Vol Sci Pap ICCAT 66(2):874–882

Melvin GD (2016) Observations of in situ Atlantic bluefin tuna (*Thunnus thynnus*) with 500-kHz multibeam sonar. ICES J Mar Sci 73(8):1975–1986. https://doi.org/10.1093/icesjms/fsw077

Pereira J (2012) Historical bluefin tuna catches from southern Portugal traps. Col Vol Sci Pap ICCAT 67:88–105

Picinetti C, Piccinetti-Manfrin G (1970) Osservazioni sulla biologia dei primi stadi giovanili del tonno (*Thunnus thynnus*, L.). Boll Pesca Piscic Idrobiol (25):223–247

Piccinetti C, Piccinetti-Manfrin G, Soro S (1997) Résultats d´une campane de recherche sur les larves de thonidés en Mediterranée. Col Vol Sci Pap ICCAT 46:207–214

Puncher GN (2015) Assessment of the population structure and temporal changes in spatial dynamics and genetic characteristics of Atlantic bluefin tuna under a fishery independent framework. Doctoral Thesis Submitted to Alma Mater Studiorum—Università di Bologna & Universiteit Gent, 237 p. http://www.iccat.int/GBYP/Documents/BIOLOGICAL%20STUDIES/Scientific_Papers/Puncher_PhD_Thesis.pdf

Puncher GN, Cariani A, Maes GE, Van Houdt J, Herten K, Cannas R et al (2018) Spatial dynamics and mixing of bluefin tuna in the Atlantic Ocean and Mediterranean Sea revéales using nex-generation sequencing. Mol Ecol Resour. https://doi.org/10.1111/1755-0998.12764

Quílez-Badia G, Cermeño P, Sainz Trápaga S, Tudela S, Di Natale A, Idrissi M, Abid N (2013) 2012 ICCAT-GBYP pop-up tagging activity, in Larache (Morocco). Col Vol Sci Pap ICCAT 69:869–877

Ravier C, Fromentin JM (2001) Long-term fluctuations in the eastern Atlantic and Mediterranean bluefin tuna population. ICES J Mar Sci 58:1299–1317

Ravier C, Fromentin JM (2004) Are long-term fluctuations in Atlantic bluefin tuna (*Thunnus thynnus*) populations related to environmental changes? Fish Oceanogr 13:145–160

Reglero P, Ciannelli L, Álvarez-Berastegui D, Balbín R, López-Jurado JL, Alemany F (2012) Geographically and environmentally driven spawning distributions of tuna species in the western Mediterranean Sea. Mar Ecol Progr Ser 463:273–284. https://doi.org/10.3354/meps09800

Restrepo VR, Díaz GA, Walter JF, Neilson J, Campana SE, Secor D, Wingate RL (2010) Updated estimate of the growth curve of western Atlantic bluefin tuna. Aquat Living Resour 23:335–342

Rey JC (1979) Interactions des populations de thon rouge (*Thunnus thynnus*) entre l´Atlantique et la Mediterranée. In: Bard FX, Le Gall JY (eds) Le thon rouge en Mediterranée: Biologie, Pêche et Aquaculture. Actes Colloq 8:87–103

Rey JC, Cort JL (1986) The tagging of the bluefin tuna (*Thunnus thynnus*) in the Mediterranean: history and analysis. CIESM, 2 p

Richardson DE, Marancik KE, Guyon JR, Lutcavage ME, Galuardi B, Lam CH, Walsh HJ, Wildes S, Yates DA, Hare JA (2016) Discovery of spawning ground reveals diverse migration strategies in Atlantic bluefin tuna (*Thunnus thynnus*). PNAS 113(12):3299–3304

Ríos Jiménez S (2007) La gran empresa almadrabero conservera andaluza entre 1919 y 1936: el nacimiento del Consorcio Nacional Almadrabero. Historia Agraria 41:57–82

Rodríguez-Marín E, Clear N, Cort JL, Megalofonou P, Neilson J, Neves dos Santos M, Olafsdottir D, Rodríguez-Cabello C, Ruiz M, Valeiras X (2007) Report of the 2006 ICCAT workshop for bluefin tuna direct ageing. Col Vol Sci Pap ICCAT 60:1349–1392

Rodríguez-Roda J (1957) El crecimiento relativo del atún (*Thunnus thynnus* (L.) de Barbate. Inv Pesq 12

Rodríguez-Roda J (1962) Talla, peso, edad y crecimiento del atún del golfo de Cádiz, España. Actas de la reunión científica mundial sobre biología del atún y especies afines. FAO Fish Rep 6 3:1823–1834

Rodríguez-Roda J (1964) Biología del atún, *Thunnus thynnus* (L.), de la costa sudatlántica española. Inv Pesq 25:33–146

Rodríguez-Roda J (1967) Fecundidad de atún, *Thunnus thynnus* (L.), de la costa sudatlántica española. Inv Pesq 31(1):33–52

Rodríguez-Roda J (1969a) Resultados de nuestras marcaciones de atunes en el golfo de Cádiz durante los años 1960–1967. Publ Tec Junt Est Pesca 8:153–158

Rodríguez-Roda J (1969b) Los atunes jóvenes y el problema de sus capturas masivas. Publicaciones Técnicas de la Junta de Estudios de Pesca. Subsecretaría de la Marina Mercante, 8:159–162

Rodríguez-Roda J (1978) Rendimiento de las almadrabas del sur de España durante los años 1962–1977, en la pesca del atún, *Thunnus thynnus* (L.). Inv Pesq 42(2):443–454

Rodríguez-Roda J (1980) Description of the Spanish bluefin (*Thunnus thynnus*) trap fishery. Col Vol Sci Pap ICCAT 11:180–183

Rooker J, Alvarado J, Block B, Dewar H, De Metrio G, Prince E, Rodríguez-Marín E, Secor D (2007) Life and stock structure of Atlantic bluefin tuna (*Thunnus thynnus*). Rev Fish Sci 15:265–310

Rooker J, Arrizabalaga H, Fraile I, Secor DH, Dettman DL, Abid N, Addis P, Deguara S, Karakulak FS, Kimoto A, Sakai O, Macias D, Santos MN (2014) Crossing the line: migratory and homing behaviours of Atlantic bluefin tuna. Mar Ecol Prog Ser 504:265–276

Ross M (2010) The glory days of the giant Scarborough tunny. British Library Cataloguing in Publication Data, 390 p. ISBN 978-0-9566375-0-5

Rouyer T, Brisset B, Bonhommeau S, Fromentin JM (2018) Update of the abundance index for juvenile fish derived from aerial surveys of bluefin tuna in the western Mediterranean Sea. Col Vol Sci Pap ICCAT 74(6):2887–2902

Santamaria N, Acone F, Di Summa A, Gentile R, Deflorio M, De Metrio G (2003) Età ed accrescimento di giovanili di tonno rosso (*Thunnus thynnus* L. 1758) nei mari meriodionali d´Italia. Biol Mar Medit 10(2):900–903

Sáñez A (1791) Diccionario histórico de las artes de pesca nacional por el Comisario Real de Guerra de Marina, 1791–1795, Madrid, 1988

Sarà G, Sarà R (2007) Feeding habits and trophic levels of bluefin tuna *Thunnus thynnus* of different size classes in the Mediterranean Sea. J Appl Ichthyol 23(2):122–127

Sella M (1929) Migrazioni e habitat del tonno (*Thunnus thynnus*) studiati col metodo degli ami, con osservazioni ull'accrescimento, sul regime delle tonnare, ecc. Memorie, R. Comitato Talassografico Italiano 159:1–24

Sorell JM, Varela JL, Goñi N, Macías D, Arrizabalaga H, Medina A (2017) Diet and consumption rate of Atlantic bluefin tuna (*Thunnus thynnus*) in the Strait of Gibraltar. Fish Res 188:112–120

Suzuki Z, Kai M (2012) Movement of Atlantic bluefin tuna toward the Strait of Gibraltar inferred from Japanese longline data. Col Vol Sci Pap ICCAT 67:322–330

Tangen M (1999) Størjefisket på vestlandet. Eide forlag 1.3.3 Bergen. Trykk og innbinding: Valdres Trykkeri. Repro Øyvind Tangen. ISBN 82-514-0586-6

Tangen M (2009) The Norwegian fishery for Atlantic bluefin tuna. Col Vol Sci Pap ICCAT 63:79–93

Tensek S, Pagá García A, Di Natale A (2018) ICCAT GBYP tagging activities in phase 6. Col Vol Sci Pap ICCAT 74(6):2861–2872

Tiews K (1978) On the disappearance of bluefin tuna in the North Sea and its ecological implications for herring and mackerel. Rapp P-v Reun Cons Int Expl Mer 172:301–309

Tsuji S, Nishikawa Y, Segawa K, Hiroe Y (1997) Distribution and abundance of *Thunnus* larvae and their relation to the oceanographic condition in the Gulf of Mexico and the Mediterranean Sea during May through August of 1994 (Draft). Col Vol Sci Pap ICCAT 46:161–176

Vanderlaan A, Jech JM, Weber TC, Rzhanov Y, Lutcavage M (2014) Direct assessment of juveniles Atlantic bluefin tuna: integrating sonar and aerial results in support of fishery independent surveys. Col Vol Sci Pap ICCAT 71(4):1617–1625

Vilela H, Cadima E (1961) Études sur les thons, *Thunnus thynnus* (L.) de la côte Portugaise. Ann Biol Cons Int Expl Mer 16:241–245

Wilson SG, Block B (2009) Habitat use in Atlantic bluefin tuna *Thunnus thynnus* inferred from diving behaviour. Endang Species Res 10:355–367

Chapter 6
Bluefin Tuna Fishing in the Bay of Biscay

Abstract The traditional fishing method for catching bluefin tuna in the Bay of Biscay used to be trolling, but in 1949 rod and live bait (bait boat) was introduced, which meant a great leap forward in the catches of this species. The different phases of fishing with this new system are briefly described as well how the fleet developed from the middle of the 20th century and the fishing seasons and type of fishes caught in this fishery.

6.1 History

Up until the middle of the 20th century ABFT was mainly caught in the Bay of Biscay by trolling (Fig. 6.1), but from 1947 a great transformation took place when the first test was made with bait boat thanks to the initiative of the shipowners of St-Jean-de-Luz (France), G. Pommereau and A. Elissalt, who had seen this fishing modality used by Japanese and North American fishermen in the Pacific Ocean targeting tuna (De la Tourrase 1951).

The Basque fishing vessels *Marie Elisabeth* and *La Nivelle*, based at the port of Ciboure (France), were the first to try out ABFT fishing with rods and live bait in 1947 (Anonymous 2008). This new system was an important development in the region as it significantly increased the possibilities of exploiting ABFT, making it possible to catch even the largest fishes using rods, and doing so in great quantities.

This new way of catching ABFT quickly crossed the border and by 1949 Spanish fishermen had already adopted it. At first wooden bait tanks were installed on deck. The water was renewed constantly (8 times/h) by a pump driven by the ship's engine. By 1954 all of the bait tanks were metallic and were built inside the hull to form part of the structure of the vessel and had a circulation system independent of the main engine.

The fleet targeting ABFT in the 1950s was made up of around 120 vessels, from S. Jean de Luz, Hendaye and Fuenterrabía (*Hondarribia*) although vessels from other ports dedicated to albacore tuna, *Thunnus alalunga* (Bonn.), also caught ABFT (De la Tourrase 1951); hence, in that decade record catches were reached, the consequences of which are studied in later chapters.

J. L. Cort and P. Abaunza, *The Bluefin Tuna Fishery in the Bay of Biscay*,
SpringerBriefs in Biology, https://doi.org/10.1007/978-3-030-11545-6_6

Fig. 6.1 Fishing with trolling An interpretation by Cort (2007) *Artist*: Lineke Zubieta (Santander, Spain) (Documentary archive, *IEO*)

In 1957 and 1958 the first echo sounders were installed, but the greatest advance came in the 1970s when most vessels installed radar and sonar. This latter provided great advantages to surface fishing.

In the 1960s the bait boat fleet of both Spain and France was made up of around 70 vessels; by the 1990s there were barely 20 (Cort 1990). There are now 15 modern vessels remaining in Spain (Fig. 6.2), equipped with the latest navigation and fishing equipments (Santiago et al. 2012). In France, however, just one remains, as bait boat was replaced by pelagic trawl from the 1990s. It must also be remembered that in the coastal ports of the Cantanbrian Sea that target albacore tuna, ABFT is also taken as a by-catch.

6.2 The Fishing Itself

The first operation consists of catching the live bait, which is done using purse seine nets. The bait species collected are mainly horse mackerel, *Trachurus trachurus* (L.); bogue, *Boops boops* (L.); sardine, *Sardina pilchardus* (Walbaum), and anchovy, *Engraulis encrasicholus* (L.). Fishes are put into the tanks (Fig. 6.3), which are filled before ABFT fishing begins.

Fig. 6.2 *Madre Guadalupe*, modern fishing vessel in *Hondarribia* (2009) (Documentary archive, *IEO*)

Fig. 6.3 Tank with live bait (http://www.photolib.noaa.gov/bigs/fish2034.jpg) (Documentary archive, *IEO*)

Fig. 6.4 Fishing for bluefin tuna using pole and live bait. Bay of Biscay (1958) (Photograph, E. Ithurria) (http://www.photolib.noaa.gov/bigs/fish2083.jpg) (Documentary archive, *IEO*)

When the bait boat fishery began in the 1950s fish were caught using rods (Fig. 6.4) or reels (Figs. 6.5a, b). Reels were used mainly in the fishing season of large specimens, which were caught one by one with the boat stationary. Once the bank of these fishes had been located the boat stopped over it and by baiting it fishing could continue for one or several days until the hold was full. Lateral watering was used in this fishery and the live bait was thrown into attract the ABFT.

Fig. 6.5 (**a**) Reel (http://www.photolib.noaa.gov/bigs/fish2079.jpg) (Documentary archive, *IEO*). (**b**) Fishing with reel and live bait (Documentary archive, *IEO*)

Fig. 6.6 Catching a small bluefin tuna spawner (> 40 kg) using a pole and live bait (2009) (Documentary archive, *IEO*)

Fig. 6.7 Spanish bait boat fishing bluefin tuna in the Bay of Biscay (1972) (http://www.photolib.noaa.gov/bigs/fish2063.jpg) (Documentary archive, *IEO*)

Over time this fishing method disappeared and nowadays whenever these spawners are targeted rods are used with the boat moving forward (Fig. 6.6).

The Bay of Biscay is an area of trophic concentration of ABFT. It is a seasonal fishery that lasts from June to October and is made up of juveniles aged 1–4 (5–30 kg) and small spawners of 5–10 years (40–150 kg) whose stay is generally shorter than the juveniles. They are more common in July and August (Fig. 6.7).

References

Anonymous (2008) La revolution du port, le thon à la canne. Altxa Mutillak 9(10):28–31

Cort JL (1990) Biología y pesca del atún rojo, *Thunnus thynnus* (L.), del mar Cantábrico. Doctoral thesis. Publicaciones especiales, IEO, vol 4, 272 pp.

Cort JL (2007) El enigma del atún rojo reproductor del Atlántico nororiental. Modalidad, Sosteni bilidad. Octavo concurso nacional de Ciencia en Acción. José L. Cort (coordinador). Instituto Español de Oceanografía. Depósito legal: SA. 538-2007, 62 pp.

De la Tourrase G (1951) La pêche aux thons sur la côte basque française et son evolution récente. Rev Trav Off Pêches Mari Tome 18, fas. 1, 66, 42 pp.

Santiago JH, Arrizabalaga, Ortiz M (2012) Standardized CPUE index of the Bay of Biscay baitboat fishery (1952–2011). ICCAT-SCRS/2012/100

Chapter 7
Research Related to Bluefin Tuna Fishing in the Bay of Biscay

Abstract This chapter reviews the research made on bluefin tuna in the Bay of Biscay fishery from a historical point of view. The subjects dealt with range from geographical origin, migrations from this area, conventional and electronic tagging, growth, and a compendium of studies on this fishery, are discussed. The ICCAT official statistics corresponding to landings, biological samplings (before and after the implementation of the PARP), the demographic structure of catches, and the abundance index of juvenile fishes are also considered.

7.1 Historical Antecedents

The first reference concerning the biology of Cantabrian Sea ABFT came from the French scientist (Heldt 1943), who deduced the migrations of the species from the Cantabrian Sea to the Mediterranean by examining hooks found in ABFT (fishes found in the fisheries of Sardinia, Sicily and Tunisia).

The first studies on ABFT biology and fishing in the Cantabrian Sea were made by the Spanish scientist Navaz (1950), who presented landing statistics for this species and described ABFT seasonality in the area, revealing that 90% of catches are taken between May and August. He also made a biometric study of specimens landed in the port of San Sebastian (Spain). His French contemporary Le Gall (1950, 1951) presented statistical data and biological samplings of French ABFT catches near the coast of Saint Jean de Luz (France), describing the seasonality of the fishery and the presence of some specimens in their period of sexual maturity, detailing the dynamic of the different tunas in the Bay of Biscay according to their weight composition. De la Tourrase (1951) carried out a large study describing ABFT fishing in the Cantabrian Sea as well as the introduction of the bait boat fishing system, with which very high fishing yields were obtained. Similarly, Creac'h (1952) studied the yield and economics of ABFT fishing in the Cantabrian Sea, providing data on the dynamics of the different groups depending on length and time of appearance in the fishing area between April and December. According to this author the number of vessels targeting tuna (ABFT and albacore tuna) in the port of Saint Jean de Luz was 85. In Pommereau (1955) the evolution of ABFT fishing on the Basque coast

J. L. Cort and P. Abaunza, *The Bluefin Tuna Fishery in the Bay of Biscay*, SpringerBriefs in Biology, https://doi.org/10.1007/978-3-030-11545-6_7

from ancient times up until the introduction of bait boat in 1948 is examined, placing emphasis on the great increase in catches that it brought with it. He also presented tuna landing statistics in Saint Jean de Luz between 1948–1954, emphasizing the maximum of 3600 t of ABFT and 1060 t of albacore tuna in 1951 and describing the characteristics of the 75 tuna fishing boats of the port.

In 1961 within the International *Council for the Exploration of the Sea* (ICES), the *Bluefin Tuna Working Group* (BFTWG) was founded, whose main mission was to compile ABFT catches within the area of that organization in order to study the relationships among the catches of the different sub-areas and the time when the tunas reached the fisheries. The scientists Hamre (Norway) and Tiews (West Germany) were elected as the first members of this group and scientists from other countries involved in ABFT fisheries were later incorporated, Aloncle and Maurin from France, and Lozano and Rodríguez-Roda from Spain. Prior to 1972, the year in which the ongoing collection of information on the ABFT fishery in the Cantabrian Sea began, the group had published five reports. In none of them had the information concerning Spanish catches in the Cantabrian Sea yet appeared, the report corresponding to the period 1973–1975 being the one containing the first Spanish data (Aloncle et al. 1977). The last report of the BFTWG was published in 1981 (Aloncle et al. 1981). At the beginning of the 1980s the BFTWG was disbanded due to the activity developing in the field of research and fisheries statistics of tunas pursued by ICCAT from the beginning of the 1970s.

From 1972, Bard et al. (1973) began observing the French and Spanish ABFT fisheries in the Bay of Biscay and soon afterwards Cort and Cendrero (1975), Cort (1976) initiated the information and sampling network that the Spanish Institute of Oceanography (*IEO*) still maintains. At the end of the 1980s the scientists of the Basque organization *AZTI-Tecnalia* joined the investigation of tuna.

7.2 Birth and Geographical Origin of Bluefin Tuna in the Bay of Biscay

In a recent study by Fraile et al. (2014) using proportions of carbon and oxygen isotopes ($\delta^{13}C$ and $\delta^{18}O$) in ABFT from the Bay of Biscay collected between 2009 and 2011, the overall contribution of ABFT from the western part of the Atlantic to the Bay of Biscay fishery was found to be <1% in those years, though in certain years this percentage can increase to 2.7%, which suggests that the fishery is almost exclusively made up of fishes born in the spawning area of the Mediterranean. In order to carry out the study the values obtained were compared with reference samples of ABFT, one from the Mediterranean and another from the Gulf of Mexico. This study confirms that in certain years there is a presence of fishes from the western Atlantic in the Bay of Biscay, just as had previously been determined by tagging (Mather III et al. 1967; Cort 1990).

Based on data from logbooks of the 1980s and publications of the time, Cort and Rey (1984) presented maps relating ABFT juvenile migrations from wintering areas and vice versa with water surface temperature. Following the same methodology, Cort (1990, 2009) applied it to the Bay of Biscay fishery. In this way, during the intermediate phase of the boreal spring from May, the 17–20 °C isotherms move northwards leading the ABFT to emigrate towards more northern areas of the ocean. Once they reach the Bay of Biscay the shoals remain in the southeastern part where the waters of the eastern edge are warmer than in the rest (García-Soto 2006). It is precisely in the oceanic part of this area that they group throughout the summer. The displacement of the isotherms southwards in October marks the start of the return of the tunas to their wintering areas. That is to say that the migration of the shoals of juvenile ABFT from wintering areas to the Bay of Biscay and vice versa is associated with the latitudinal displacement of the surface isotherms. Nevertheless, Arregui et al. (2018) show that this association is not entirely accurate, as some juveniles carrying electronic tags have occupied winter habitats in which water temperatures can reach 10 °C, something that was also described in Cort et al. (2014).

7.3 Migrations in and from the Bay of Biscay

The arrival of juvenile shoals in the Bay of Biscay usually takes place at the end of the boreal spring (May–June), mainly from the west of the Iberian Peninsula, in some cases around the cold waters of the Portuguese continental shelf (Gil 2006) and in others from the northwest and mid-Atlantic or from the vicinity of the Bay of Biscay (Goñi et al. 2010; Arregui et al. 2018), their final destination being the bottom of the Bay where they find an abundance of food (Ortiz de Zárate and Cort 1986; Logan et al. 2010a, b). In the interior of the Bay the tunas form large shoals of juvenile fishes of the same age (Cort 1990) and search for food continually. These shoals act differently depending on the nature and behaviour of their preys. Their formations and the effects they bring about on the sea surface give away their presence, which provides fishermen with the opportunity to make catches.

The small spawners (up to 2.2 m) arrive in mid-July and by the end of August they can hardly be found in this area (Cort 1976; Cort and Rodríguez-Marín 2009), although in recent years they are frequent in autumn (October). The recovery of a juvenile ABFT tagged in the Bay of Biscay (1990) and recovered in the North Sea (60° N/8° W) eight years later (Cort unpublished) as well as the presence of these fishes to the south of 62° N, caught by Norwegian vessels in August (Hamre and Tiews 1963; Tangen 2009) are evidence of the short seasonality of this group in the Bay of Biscay.

7.4 Bluefin Tuna Migrations to and from the Bay of Biscay. Conventional Tagging

Cort et al. (2010) describe the different types of tags used in ABFT tagging, and establish as conventional tags those that are made up of small plastic tubes (2 mm width and 12 cm length) with a little harpoon on the end, which are inserted in the dorsal part of the fish by means of a steel applicator.

Following a description of the juvenile ABFT of the Atlantic coasts of Morocco made by Furnestin and Dardignac (1962), in the 1970s French scientists tagged ABFT juveniles in that area, some of which were recovered in the Bay of Biscay in the following and successive years (Aloncle 1973; Lamboeuf 1975; Brêthes 1978, 1979). It was the first time ABFT tagged in other areas of the eastern stock had been recovered in the Bay of Biscay. Years later, and as a result of tagging by Spanish scientists on Spanish Mediterranean coasts, some of those fishes were also recovered in the Bay of Biscay (Cort 1990), which shows the interchange of ABFT of the same stock among different areas.

French scientists had tagged 34 ABFT in the westernmost part of the Cantabrian Sea during the albacore tuna tagging surveys between 1967 and 1972, of which two were recovered in the western Atlantic (Aloncle 1973). This demonstrated that juvenile ABFT make transatlantic migrations from East to West, something that had already been shown in the other direction, from West to East, by Mather III et al. (1967). The surveys targeting ABFT organized by the *IEO* began in 1978 and continued until 1991. A total of 5,653 ABFT were tagged, mainly juveniles, of which 360 (6.4%) were recovered in the following years: 329 (91.4%) turned up in the Bay of Biscay again; 16 (4.4%) were recovered in the Atlantic fisheries of the U.S.A. (Table 7.1); 7 (1.9%) in the Mediterranean, and 8 (2.2%) in diverse eastern Atlantic fisheries. The ABFT that were recovered in the fisheries of the eastern Atlantic and Mediterranean were adults and some of them, which had spent up to 9 years at liberty, had increased in weight by up to 250 kg (Cort 2006).

The surveys were carried out on board commercial fishing vessels using troll in the first years and later, from 1979, on bait boat vessels. The tagging system was always the same: the fishes were taken to the tagging cradles, where they were measured and tagged with conventional tags. Once tagged, they were freed in an operation that lasted a few seconds.

The results of these experiments show that most of the juveniles tagged in the Bay of Biscay, and while they are still juveniles, return in later years which has been shown clearly by Arregui et al. (2018), who refer to their "strong fidelity to the area". Having reached sexual maturity, fishes tagged in the Bay of Biscay have been recovered in areas in which spawners are common, as shown in Fig. 7.1. This demonstrates the connection of ABFT with other fisheries of the eastern and western stocks (Cort 1990, 2006). With the exception of the recoveries that took place in the year following tagging, above all those recovered in the Gulf of Lion a year afterwards, we know nothing of their movements during the years they were at

Table 7.1 Count of bluefin tunas making transatlantic migrations, tagged in *IEO* surveys in the Bay of Biscay (1978–1990)

Tagging						Recapture				
Tag number	Date (month, day, year)	Latitude	Longitude	*FL* (cm)	Age (years)	Date (month, day, year)	Latitude	Longitude	*FL* (cm)	*RW* (kg)
R 3874	08-20-1978	43.50 N	2.35 W	78	2	01-02-1980	42.18 N	60.45 W	–	12
R 7336	09-13-1979	44.20 N	2.40 W	103	3	08-25-1980	43.00 N	69.00 W	114	27
R 7388	09-13-1979	44.20 N	2.40 W	85	2	12-15-1985	41.02 N	51.01 W	202	150
R 9706	08-17-1980	43.40 N	3.15 W	60	1	08-13-1982	40.36 N	72.03 W	112	–
R 9757	08-17-1980	43.40 N	3.15 W	84	2	09-05-1981	39.40 N	72.40 W	93	14
S 2469	08-04-1980	43.55 N	3.03 W	80	2	08-10-1981	39.40 N	72.40 W	99	–
S 5898	08-13-1982	44.30 N	2.25 W	60	1	09-07-1983	41.30 N	72.30 W	94	18
AT 3547	08-09-1985	43.30 N	1.48 W	80	2	08-25-1988	45.00 N	66.00 W	–	–
AT 3869	09-30-1986	43.43 N	2.55 W	64	1	06-25-1988	38.50 N	73.58 W	114	32
EM 7172	10-05-1986	43.57 N	2.27 W	66	1	08-15-1989	39.44 N	73.20 W	130	–
EM 7218	10-05-1986	43.57 N	2.27 W	67	1	09-03-1987	40.04 N	73.40 W	105	19
EM 7486	10-07-1986	43.55 N	2.31 W	64	1	08-13-1988	39.20 N	73.45 W	100	18
EM 8479	08-03-1988	44.18 N	2.24 W	60	1	10-15-1989	38.00 N	75.00 W	90	–
YF 915	08-25-1990	44.00 N	3.37 W	64	1	09-29-1991	40.33 N	71.28 W	92	16
YF 985	08-25-1990	44.00 N	3.37 W	65	1	08-28-1991	40.45 N	71.52 W	91	16
YF 6153	08-20-1990	43.13 N	3.21 W	63	1	07-10-1993	36.45 N	75.10 W	125	35

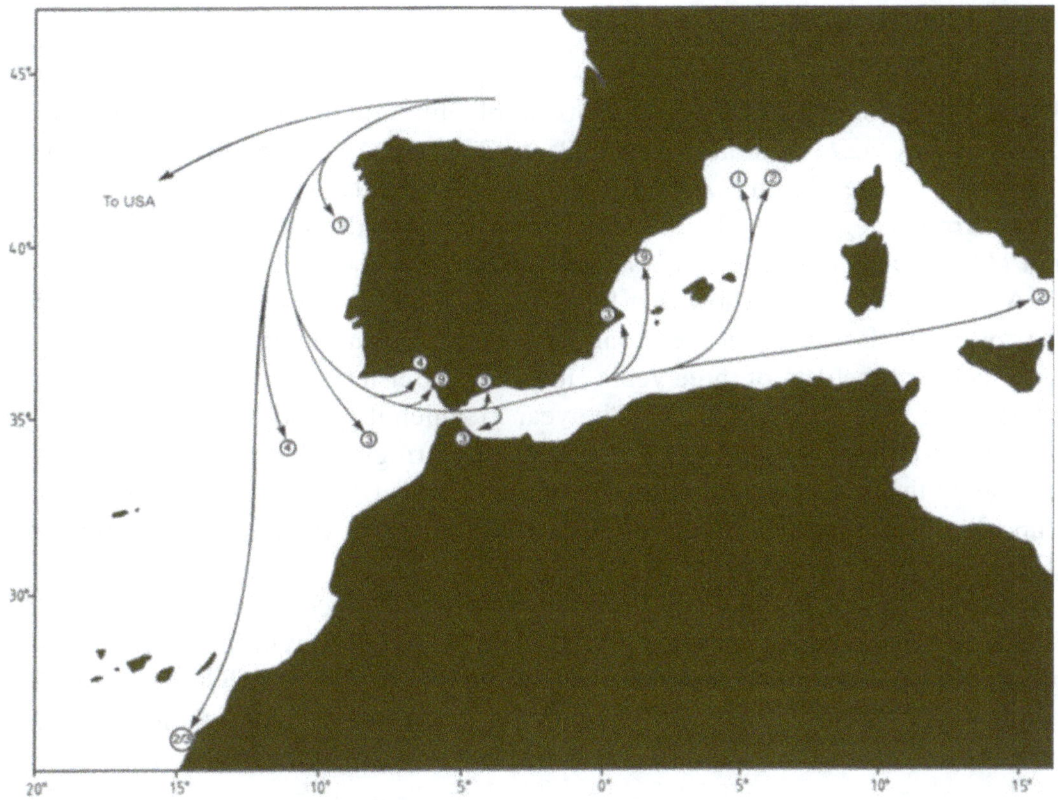

Fig. 7.1 Some of the long migrations of young bluefin tunas tagged in *IEO* surveys in the Bay of Biscay (1978–1990) (The numbers in the circles indicate years at liberty. A reference is included to the 17 specimens that emigrated to the coasts of the U.S.A.). Taken from Cort (2006)

liberty. It is particularly interesting to note that migrations are made by juveniles to the western Atlantic from the Bay of Biscay fishery.

Nowadays, with electronic tags that tell us practically everything about the movements of the fish and its environment, conventional tagging has lost its importance. Even so, other conventional tagging surveys have been conducted, such as those of 2005 and 2007, organized by *AZTI-Tecnalia* and the *IEO*, in which 2,141 were tagged (Rodríguez-Marín et al. 2008a). Later, *AZTI-Tecnalia* tagged 7,698 bluefin tunas in a survey organized by ICCAT-GBYP between 2012 and 2015. The results of these surveys (Tensek et al. 2018) confirm those from surveys carried out in the past. Overall, the Bay of Biscay is the area in which most eastern stock bluefin tunas have been tagged.

In addition to the information obtained from the recovered fishes regarding migrations, many of them have been used to carry out growth analysis, all of which comes thanks to the fact that they were measured when tagged and the fishermen who recovered them facilitated data on their sizes, though this does not always happen.

7.5 Electronic Tagging

The first electronic tagging experiments on bluefin tuna in the Bay of Biscay were conducted by *AZTI-Tecnalia* from 2005 to 2009, when 136 archival tags and 29 pop-up tags were placed ($Lh = 60–107$ cm) (Arrizabalaga et al. 2008; Goñi et al. 2010; Arregui et al. 2018). In 2009 the University of Cádiz and IEO tagged a further 101 ($Lh = 60–85$ cm) (Medina et al. 2011).

Arregui et al. (2018) show that five of the internal tags were recovered, four in the following 1–5 years in the Atlantic and one 7 years later. This latter tag had only recorded the data of three years and when the fish was recovered it was in the adult phase in the Mediterranean. The year in which it passed through the Strait of Gibraltar for the first time could not be known (Arregui, pers. comm.). The results also show that the juvenile wintering areas are found between the Bay of Biscay, Cape San Vicente and the Azores. Similarly, the western Atlantic is a wintering area (Cort et al. 2014; Arregui et al. 2018).

The recovery rate of internal electronic tags has fallen considerably with respect to that of conventional tags since these activities coincided with the start of PRPA, which has led to a considerable fall in fishing activity in this area. ICCAT-GBYP has subsequently organized diverse electronic tagging surveys, one of them in the Bay of Biscay. Tensek et al. (2017) cite and reveal the results of these activities.

7.6 Age and Growth of Bluefin Tuna in the Bay of Biscay. Age-Length Key and Length-Weight Relationship

The first scientists to publish ABFT growth studies in the Bay of Biscay were Compeán-Jiménez and Bard (1983). They are credited with the use of fin rays for ageing, a technique that had not previously been used for this kind of study in ABFT. The methodology represented a large step forward from the one in use up until then (otolith age-reading), given that the fish does not get damaged and so samples are obtained at no cost since the sale of the fish is not affected. Some years later, using the same methodology, Cort (1990) specified some aspects regarding observable bands in the samples and presented an integral growth equation through a combination of data from the Bay of Biscay (fishes up to 200 cm) and samples from the traps fishery of the Bay of Cadiz (192 fin rays from fishes up to 3 m caught in June, 1984). Age groups of juveniles and small adults (ages 1–8 years) of the Bay of Biscay and adults of 6 to 19 years from the Bay of Cadiz were combined. The full equation that resulted, using the classical model of Von Bertalanffy (1938) was:

$$L_t = 318.85\left[1 - e^{-0.093(t+0.97)}\right]$$

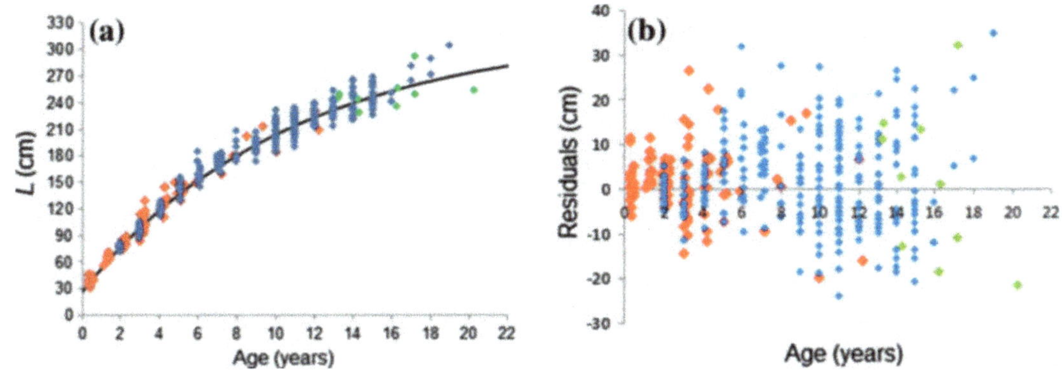

Fig. 7.2 *Left*: Growth curve (*solid line*) and readings/age of fin rays (*blue points*) and growth data from tagging-recovery (*red points*, eastern Atlantic; *green points*, western Atlantic). *Right*: Residuals against age. Taken from Cort et al. (2014) (Courtesy of *Taylor and Francis Group*)

where: L_t = fish length; L_∞ = asymptotic value of the growth curve, supposeddy very close to $L_{\max}$ (máximum population length); k = growth rate (annual); t = time (age, generally in years); t_0 = theoretical instant when $L_t = 0$.

This equation was adopted by the ABFT assessment group in 1991 and remains a reference today. It is still in use by this group for eastern stock assessments. For its validation Cort et al. (2014) used different methodologies, one of which was to superimpose the age readings of 578 ABFT caught in the Bay of Biscay and Mediterranean Sea onto the growth curve, as well as 131 data obtained from recaptured ABFT, fundamentally from the Bay of Biscay. The result of the analysis indicated a good fit of the data to the model.

On the left of Fig. 7.2 the growth curve of the eastern stock is shown (solid line) on which readings/age of fin rays (blue points) and growth data from tagging-recovery (red points, eastern Atlantic; green points, western Atlantic) have been superimposed. On the right are the corresponding residuals.

The only way to evaluate the predictive power of the curve is by observing a posteriori and analyzing the residuals. The analysis of these indicates that there are no systematic errors in the model as it presents a symmetrical distribution of the data on both sides of the x-axis. The values of the analysis are given in Cort et al. (2014).

A few years after spine reading for age determination began, the first age-length key (ALK) for the Bay of Biscay fishery was established (Rey and Cort 1984; Cort 1990).

The Bay of Biscay fishery is made up of juveniles aged 1–4 (5–30 kg) and of small spawners of 5 to >10 years (40–150 kg). Usually, the juveniles remain accessible to fishing from June until October–November and small spawners mainly in July and August (Cort and Nøttestad 2007; Cort and Rodríguez-Marín 2009).

The analytical expression relating length and weight is: $P = b\,L^k$, in which:

P	weight of the individual (kg)
L	length of the individual (cm)
b and k	constants.

Table 7.2 Parameters of the length-weight relationships of bluefin tuna in the Cantabrian Sea

Season	Sampled fish	*b*	*k*	Size range (cm)	r^2
Summer	173	0.00004388	2.815	60–195	0.993
Autumn	102	0.00003856	2.859	65–115	0.993

Table 7.2 from Cort (1990) shows the length-weight relationships for the seasons in which there is a greater presence of specimens: summer and boreal autumn (June–October).

In relation to larger fishes passing through the Cantabrian Sea and Bay of Biscay, the length-weight relationship in summer reflects the fact that these fishes are in still in a state of reduced fattening condition.

7.7 A Fin Ray Sample that Confirms the Nature of Visible Rings

Based on a sample of 1,536 spines collected between 1980 and 1986, Cort (1990) describes the different visible rings in fin ray cross-sections as follows:

- The hyaline rings are "winter" rings and are formed between autumn and boreal spring (November/April–May).
- The single hyaline rings may be thin or thick. If they are thin it indicates that there was considerable slowing in growth and if they are thick it means that there was more growth but with food with little protein.
- In the double hyaline rings the first ring is formed at the start of autumn and the second at the end of winter-spring.
- The opaque band (active growth) represents the growth that takes place between June–November, when food is very rich in proteins.
- The tracking of the opaque band between June and November (marginal growth) shows the exponential growth that stops in the winter months.
- At the beginning of the fishing season (June) the recently formed winter ring is at the edge of the fin ray; there is hardly any opaque band. At the end of the fishing season (October–November) the ring is distant from the edge and the opaque band is very broad.

Despite the above, the fin raysamples vary considerably, and in addition to single rings (thin and thick) and double rings, there are triple rings and false ones; singles and doubles appearing even in the same sample.

Twenty four years after making this description, the nature of the different observable parts in the cross-sections was confirmed in a fin ray recovered from an ABFT tagged with an internal electronic tag, from which all movements of the fish are known over the two years that the migration lasted between the fish being caught,

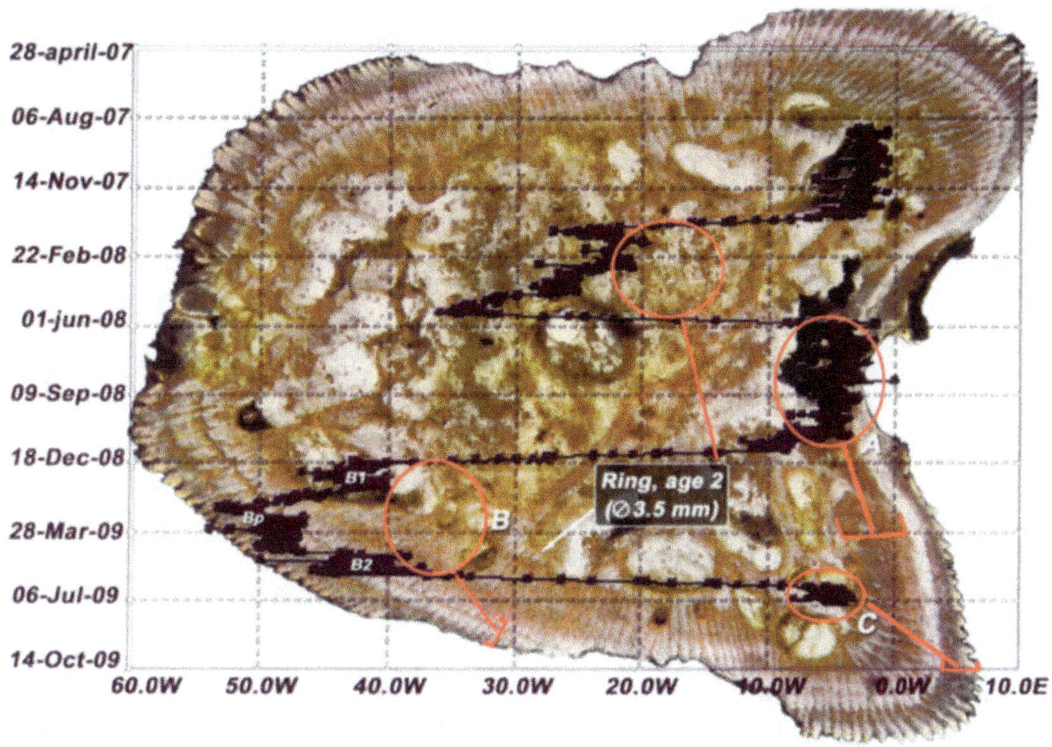

Fig. 7.3 Transversal cross-section of the spine of the first dorsal fin of a bluefin tuna tagged and recovered with an electronic tag. The estimated geolocation (dark-colored graph) has been superimposed (explanation in the text) Data of the fish: *Fl* = 64 cm; age: 1 year (at tagging); *Fl* = 102 cm; age: 3 years (on recovery). Ø spine = 5.3 mm (at the broadest part). Taken from Cort et al. (2014) (Courtesy of *Taylor and Francis Group*)

tagged and freed in the Bay of Biscay (18 August, 2007) and recovered in the same area two years later (2 August, 2009).

If it is difficult to recover an internal electronic tag, it is even more so and obtain the fin ray from the same fish as in this case. It is exceptional and unprecedented. The interpretation of the visible bands in the ray cross-section together with the movement of the fish (Fig. 7.3) revealed the "winter" nature of the hyaline bands described by Cort (1990), and it was also possible to confirm that the double rings (in the sample studied) were formed during the transatlantic migrations, westwards and eastwards, of the fish during the second winter after having been freed. The article that contains this information was published in *Reviews in Fisheries Science & Aquaculture* in August 2014 (Cort et al. 2014).

The interpretation of this sample, which must be made together with Figs. 7.3 and 7.4, is the following:

The movements of the fish over two years are represented on the dark-colored graph in the interior of the fin ray cross-section (Fig. 7.3) and are interpreted using the data from the two axes. Thus, for example, 14-Nov-07 indicates that the fish was still in the Bay of Biscay (longitude 2° W) after having been tagged on 18 August; on 22-Feb-08 it was wintering in the middle of the Atlantic Ocean (25° W); and on

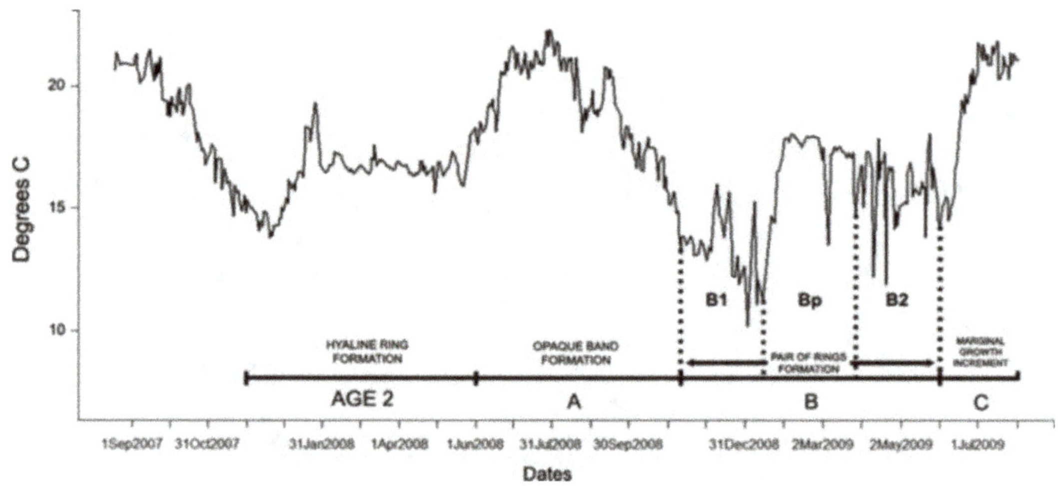

Fig. 7.4 Graph of water surface temperature recorded by the internal electronic tag indicating the moments when the hyaline rings and opaque bands are formed (Explanation in the text). Taken from Cort et al. (2014) (Courtesy of *Taylor and Francis Group*)

01-Jun-08 it had returned to the Bay of Biscay where it would remain until November 2008 (2°–8° W). On 18-Dec-08 it was wintering in the western Atlantic at longitude 45° W; on 28-Mar-09 it reached 55° W off the coasts of the U.S.A. and Canada; and finally, a very swift migration saw it return to the Bay of Biscay (06-Jul-09) where it was recovered in August 2009. According to Arregui et al. (2018), this fish crossed the Atlantic in 22 days.

Figure 7.4 shows the temperature in which the fish moved while it was at liberty. In the Bay of Biscay the water temperature went from 22 °C in August 2007 to 14 °C in December 2007, when it migrated to the mid-Atlantic. Between January and April 2008 it moved in more temperate waters (17 °C). During this time the winter ring formed as represented in Fig. 7.3 with a red circle *Ring, age* 2 (Ø 3.5 mm) and a small white arrow that points towards the ring; represented as a *HYALINE RING FORMATION* (AGE 2) in Fig. 7.4.

From this moment the letters A, B and C are added to both Figs. 7.3 and 7.4, in which the visible structures in the spine are shown. In this way, A represents the stay of the fish in the Bay of Biscay during the summer of 2008, which is when the opaque band forms in the spine (Fig. 7.3). The water temperature is similar to that of the previous summer-autumn. Area B is the second winter in which the fish migrates across the Atlantic (to 55° W) between December 2008 and May 2009. In both figures the different phases of its long stay in the northwest Atlantic are represented by the letters B_1, B_p and B_2, which is when the pair of hyaline bands forms. The values of water temperature in these dates are completely different to those of the previous winter since the fish enters and leaves the Gulf Stream (Arregui, pers. comm.; Cort et al. 2014), and hence the abrupt falls and rises in values, which vary between 10 and 17.5 °C. Lastly, area C is the marginal growth increment of the spine, the active growth of the fish when it has returned to the Bay of Biscay shortly before being recovered in August 2009.

It must be pointed out that although the latitude is not represented on the graph, the fish moved between 35° and 50° North (Arregui et al. 2018).

7.8 Other Studies on Bluefin Tuna

In the Bay of Biscay studies have also been carried out on feeding (Ortiz de Zárate and Cort 1986; Logan et al. 2010a, b), feeding patterns (Cort and Rey 1979), microelements in otoliths (Rooker et al. 2014; Fraile et al. 2014), parasites (Cort 1990; Rodríguez-Marín et al. 2008b), reproduction (Cort et al. 1976), biometry (Cort 1990), comparative morphology (Addis et al. 2014), behaviour in feeding and regarding predators (Cort 1990), age composition of shoals (Cort 1990), presence of large specimens (Cort 1980, 1981); population dynamics (Bard and Cort 1979, 1980; Cort 1990), and the description of fishing systems (Merino 1997). Similarly, in the context of the biological and genetic sampling and analyses by the program ICCAT-GBYP, the Bay of Biscay has been an important sampling area (Di Natale et al. 2018).

7.9 Official Bluefin Tuna Catches in the Eastern Atlantic (ICCAT, 1950–2016)

The Fig. 7.5 is an advance of the results of the analysis made in the last section of the present article determining the main reasons behind the events described below.

On one hand, ABFT juvenile catches taken in the Bay of Biscay between 1950 and 2016 and those of Morocco (Atlantic part) between 1958 and 2004 are shown (in grey). On the other, the catches of spawners in the Atlantic area mainly caught by the traps in the Strait of Gibraltar, by Asian longliners, basically Japanese, and in the northern European fisheries mostly by Norwegian purse seiners, are represented (in black). The data are taken from the official statistics of ICCAT (ICCAT Database 2017) that appear in ICCAT (2017).

There are four distinct stages:

1950–1963: This period saw a spectacular increase in catches of spawners as a result of the introduction of ABFT fishing in the north of Europe, mainly of purse seine in Norway (Hamre 1960). At the same time the juvenile bait boat fisheries developed in the Bay of Biscay and purse seine in Morocco (Atlantic area). While higher catches were obtained in these latter fisheries, those of spawners fell drastically within a few years.

1963–1975: The fall of the spawner fisheries led to the practical disappearance of the traps in the Strait of Gibraltar and the start of the collapse of fisheries in the north of Europe, which came a few years later (Tiews 1978). Meanwhile, the juvenile fisheries remained stable.

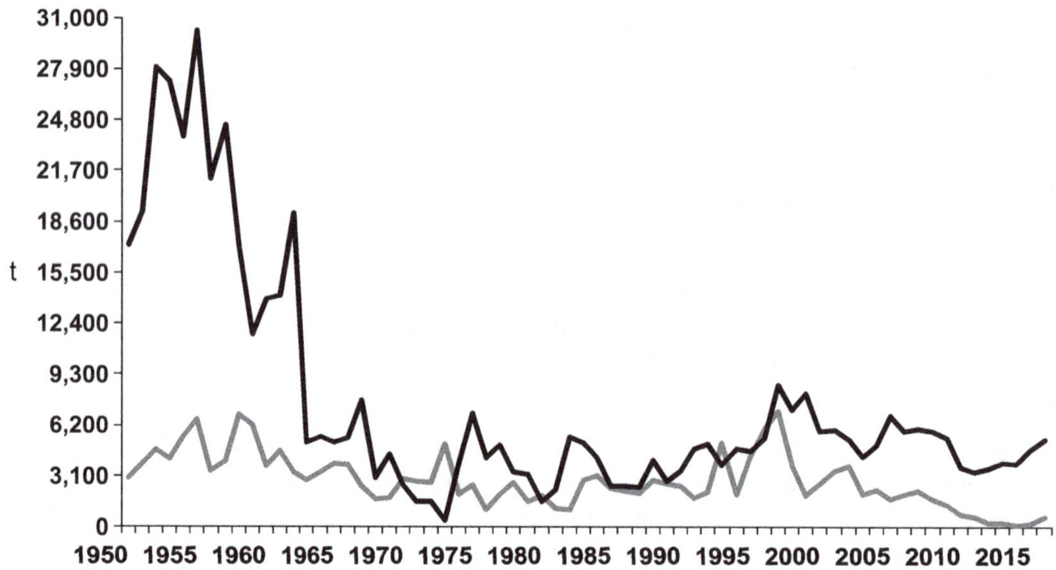

Fig. 7.5 Official statistics of bluefin tuna catches in the eastern Atlantic (metric tons). *Black line*: Catches of ABFT spawners using traps, LL, BB: *Sportive Gray line*: ABFT juvenile catches with BB in the Bay of Biscay and PS in Morocco, Atlantic side Legend: BB = Bait boat; PS = Purse seine; LL = Longline

1975–2008: The spawner fisheries of the Strait of Gibraltar recovered a little, though never to the levels of the 1950s, as a result of the installation of a few traps in the area of the Strait of Gibraltar (Cort et al. 2012), but those of northern Europe collapsed at the beginning of the 1980s. The juvenile fisheries (mainly in the Bay of Biscay) continued to operate without changes. In 2008 the PRPA of ICCAT (2008) was implemented.

2008–the present: with the PRPA fully enforced it is difficult to make any conjectures since catches only reflect what is taken in each fishery in accordance with the fishing quotas assigned. While the spawner fisheries, basically traps and longline continue and grow, the juvenile fisheries are in a drastic decline as fishermen do better economically by selling their quotas.

7.10 Historical Series of Catches in the Bay of Biscay

The ABFT fishery of the Bay of Biscay has traditionally been one of the most important of the eastern stock. As the fleets of both France and Spain have caught ABFT in this area any study must consider the contributions of both countries. In Cort and Bard (1980) this fishery is described within the framework of the SCRS.

The first historical series of ABFT catches in the Bay of Biscay (Spain + France) between 1950 and 1986 was published by the IEO in 1990 (Cort 1990). Figure 7.6

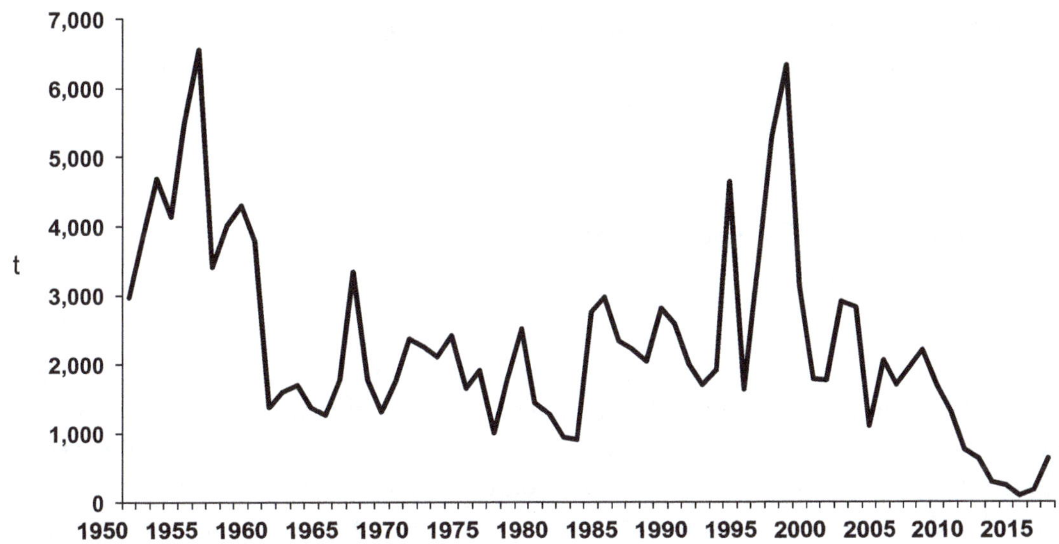

Fig. 7.6 Historical series of bluefin tuna catches (metric tons) in the Bay of Biscay (Spain+France), 1950–2016

presents this series projected onwards to the present thanks to a review of catches made by the IEO and *AZTI* (supported by ICCAT-GBYP), which ended with joint publications presented to the SCRS (Cort et al. 2015).

In the series there is a first phase (1950s) with France taking the maximum catches and dominating the fishery. Over time, its importance fell and at the end of the last century there were practically no French bait boat vessels left, the fleet having reconverted to pelagic pair trawl gear. The increase that began to appear in the 1990s was due to the expansion in fishing effort by Spain as a result of improved technologies, which were becoming more advanced every year.

In the initial phase of the fishery the mean catches of France and Spain were 2,200 and 2,100 t/year respectively, whereas towards the end of the 20th century between 1990 and 2000 an average of 3,100 t/year were caught in Spain and 560 t/year in France, confirming the decline of the French bluefin tuna fishery in the Bay of Biscay. The fall in the end phase of the series was due to the implementation of the PRPA from 2008, which assigned a quota of less than 1000 t/year to this fishery, which has not even been reached in several cases because it was sold to other fisheries, such as the traps and purse seiners.

Spanish catches by port are detailed in Cort et al. (2015), who point out that 97.1% of the Spanish bluefin tuna catches of the Bay of Biscay come from Basque ports, making up 85,613 t of the total of 88,170 t. The rest correspond to other regions such as Cantabria, Asturias and Galicia. It also states that vessels targeting this species caught 81.8%, while the remaining 18.2% corresponded to by-catches taken by other fleets, mainly those targeting *Thunnus alalunga*.

This historical series is evidently highly stable. Although there are fluctuations, the fall that took place in the middle of the 2000s was indeed drastic, just when the PRPA of ICCAT was implemented. Moreover, it must be remembered that this series is very affected in the last years by a new strategy from northern Spain, which consists of selling quotas to other fisheries, such as the Mediterranean traps and purse seine fisheries, which is cited in the paper of Tensek et al. (2018).

7.11 Biological Sampling. Demographic Structure of Catches

In the Bay of Biscay fishery ABFT length samplings have been made regularly since 1972. Though it was French scientists who began this work (Bard et al. 1973), samplings have since been made in Spain, mainly in the port of Fuenterrabía (*Hondarribia*). The measurement taken is zoological length (from the snout to the outer part of the tail fin, *Fl*), and distributions at age are drawn up using the length by applying age-length keys, the age of the fish being determined by reading fin rays. This practice is fundamental to assessments of the population since they require analytical models, which depend upon the demographic structure of catches.

Since 1950 when the French scientist J. Le Gall published the distribution of ABFT lengths in the Bay of Biscay in ICES (Le Gall 1950, 1951) we have known that this fishery has been mainly made up of juvenile fishes. In his articles this author also referred to the seasonality of spawners (>40 kg), which pass through this area between July, August and mid-September (Table 7.3).

From this sampling came the information that, of a catch of 3,759 t, 70% were juveniles aged 1–4 years. This fact has been confirmed by Bard et al. (1973), Cort (1990), and Cort and Abaunza (2015).

Figures 7.7, 7.8 and 7.9 are length samplings made before the implementation of the PARP.

In these three cases we see that of a very high total catch (between 172,000 and 353,000 fishes/year) almost all of the fishes caught (98.9%) were juveniles (<5 years). This was at a time when there were not many spawners in the Atlantic and small-sized fishes commanded a high price due to the French market.

Table 7.3 Samplings carried out in St-Jean-de-Luz (France) between 13 April and 23 October 1951. (Adapted from Le Gall 1951)

<table>
<tr><th>Year 1951</th><th>April</th><th>May</th><th>June</th><th>July</th><th>August</th><th>September</th><th>October</th></tr>
<tr><td>Catch (t)</td><td>12</td><td>258</td><td>665</td><td>652</td><td>838</td><td>867</td><td>467</td></tr>
<tr><td rowspan="2">Fork length (cm)</td><td rowspan="2">65–78</td><td rowspan="2">65–70</td><td>70–92</td><td rowspan="2">>120</td><td rowspan="2">120–145</td><td>120–145</td><td rowspan="2">76–78</td></tr>
<tr><td>95–115</td><td>68–76</td></tr>
<tr><td rowspan="2">Round weight (kg)</td><td rowspan="2">4.1–8.7</td><td rowspan="2">4.6–8.7</td><td>7–17</td><td rowspan="2">23–41</td><td rowspan="2">29–70</td><td>29–70</td><td rowspan="2">7–9</td></tr>
<tr><td>17–29</td><td>6–8</td></tr>
<tr><td rowspan="2">Age</td><td rowspan="2">1–2</td><td rowspan="2">1–2</td><td>2–3</td><td rowspan="2">3–5</td><td rowspan="2">4–6</td><td>4–8</td><td rowspan="2">1–2</td></tr>
<tr><td>3–4</td><td>1–2</td></tr>
</table>

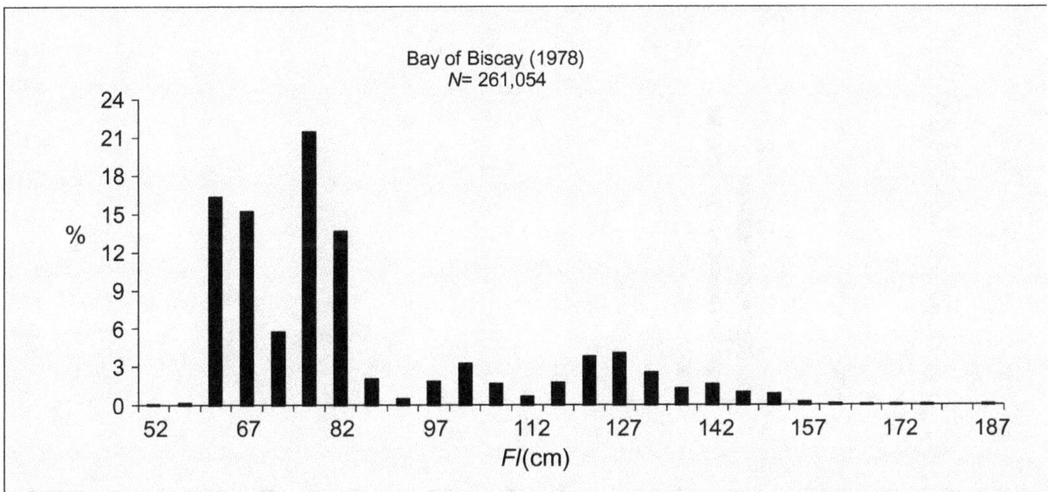

Age	1	2	3	4	5	6	7	8
(%)	63.5	25.7	3.8	5.5	1.4	0.1	0.02	0.01

Fig. 7.7 Length frequencies and corresponding ages, year 1978 (before the implementation of the PARP) Bay of Biscay N = number of fish caught (Fl = Fork length)

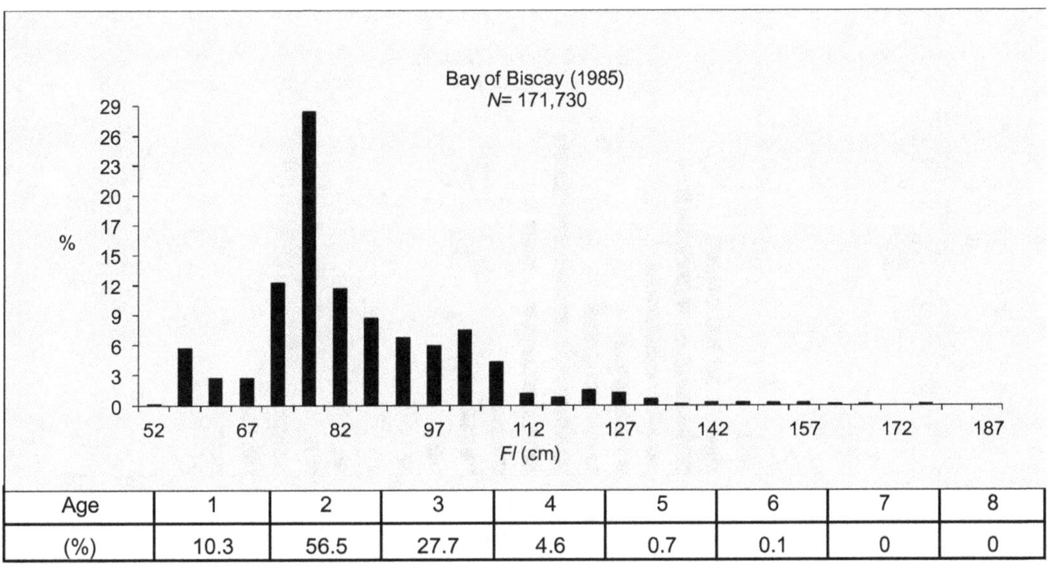

Age	1	2	3	4	5	6	7	8
(%)	10.3	56.5	27.7	4.6	0.7	0.1	0	0

Fig. 7.8 Length frequencies and corresponding ages, year 1984 (before the implementation of the PARP) Bay of Biscay N = number of fish caught (Fl = Fork length)

Since the implementation of the PARP (2008) this fishery has undergone considerable changes:

- The minimum length has been raised to 8 kg, which means that fishes aged one year and some of age two are no longer caught.
- The reduced quotas assigned in recent years (2012–2015) have led to their sale to other spawner fisheries, traps and purse seine.

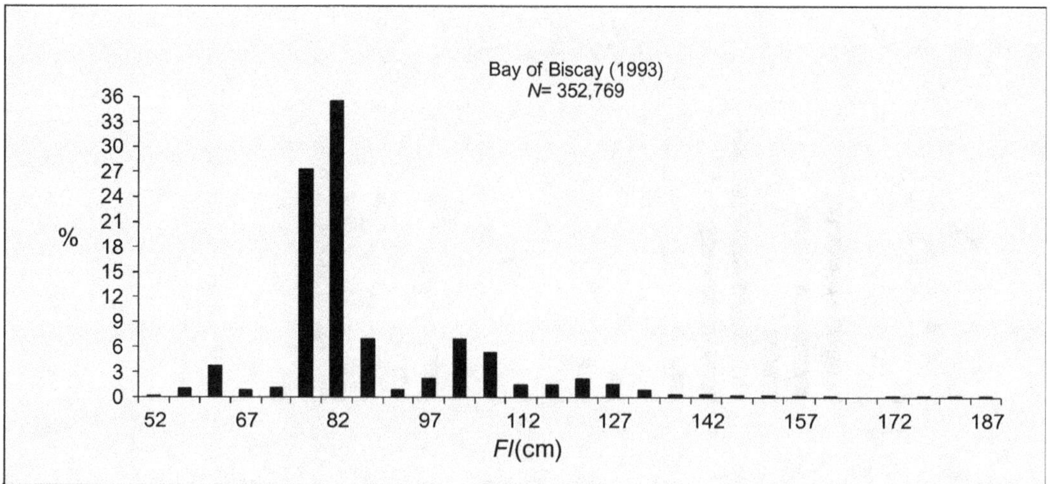

Age	1	2	3	4	5	6	7	8
%	13.5	61.5	18.9	5.2	0.8	0.1	0.002	0

Fig. 7.9 Length frequencies and corresponding ages, year 1993 (before the implementation of the PARP) Bay of Biscay N = number of fish caught (Fl = Fork length)

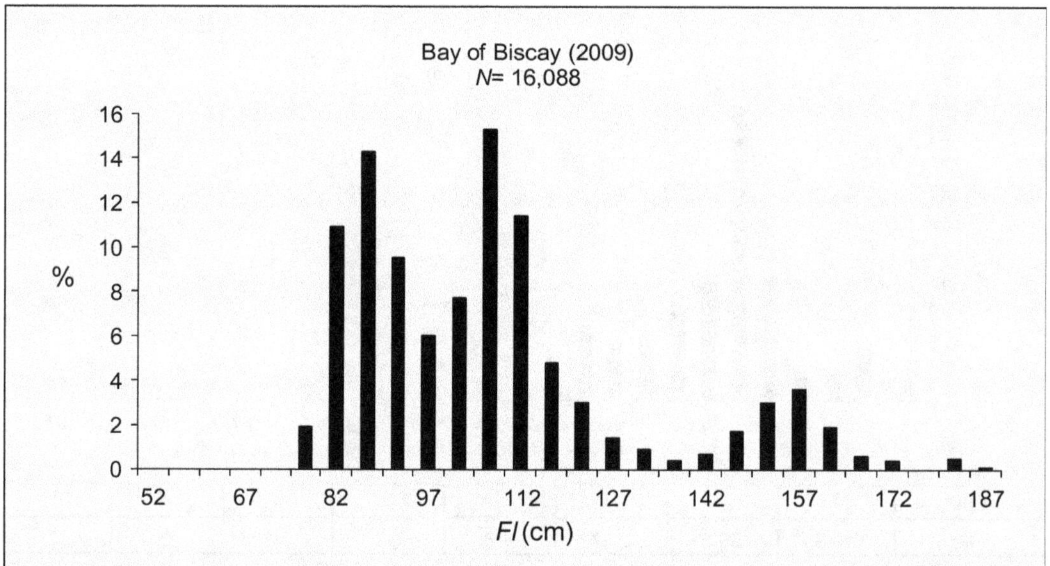

Age	1	2	3	4	5	6	7	8
%	0	27.0	38.2	21.0	8.0	5.3	0.2	0.2

Fig. 7.10 Length frequencies and corresponding ages, year 2009 (after the implementation of the PARP) Bay of Biscay N = number of fish caught (Fl = Fork length)

- In the first year of fishing (2016) after selling the quota, the fleet has targeted large-sized fishes, which fetch a better price.

These facts can be observed in the following two examples (Figs. 7.10 and 7.11) together with the total number of fishes caught, which is actually very small in comparison with previous decades.

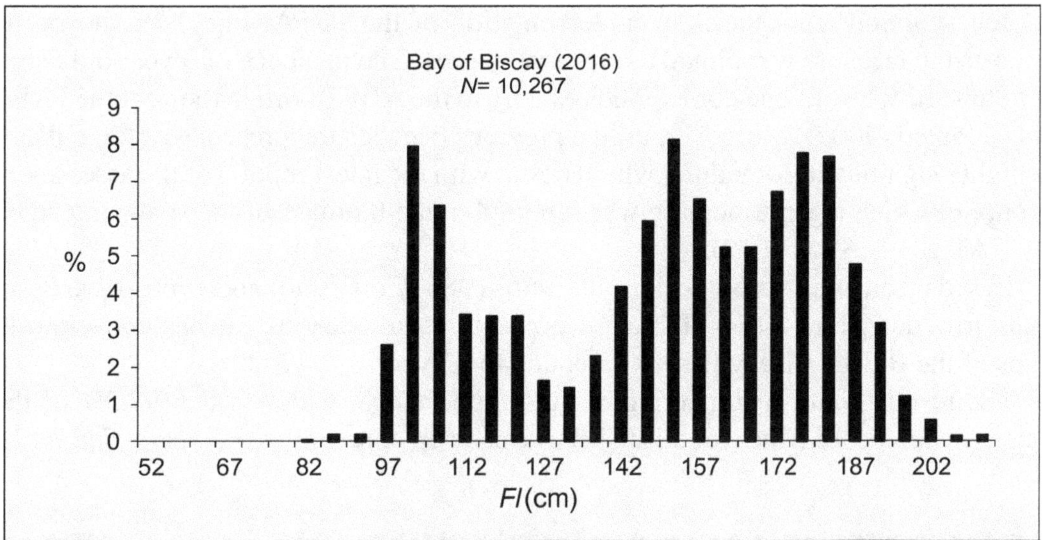

Age	1	2	3	4	5	6	7	8	9	10	>10
%	0	0.7	15.8	13.4	23.3	17.9	8.5	5.5	11.0	3.0	0.9
Cohort		2014	2013	2012	2011	2010	2009	2008	2007	2006	2005

Fig. 7.11 Length frequencies and corresponding ages, year 2016 (after the implementation of the PARP) Bay of Biscay N = number of fish caught (Fl = Fork length)

While in 2009 (Fig. 7.10) juveniles aged 2–4 (86.2%) were still being caught, the presence of spawners in the catch was clear (Cort and Martínez 2010). In 2016, following four years in which the quota was being sold off, fishing switched to large-sized fishes as this was more profitable. By that year the juvenile catch (2–4 years) had fallen to 29.9%.

Table 7.4 Demographic structure of bluefin tuna catches in the Bay of Biscay (1949–2010)

1949–2010					
Age	Mean weight (kg)	Num. Fish	Catch (kg)	% fish	% weight
1	5	4,774,272	23,871,360	37.3	15.3
2	11	5,459,690	60,056,590	42.7	38.6
3	20	1,744,950	34,899,000	13.6	22.4
4	37	524,681	19,413,197	4.1	12.5
5	53	179,399	9,508,147	1.4	6.1
6	66	63,311	4,178,526	0.5	2.7
7	80	28,954	2,316,320	0.2	1.5
8	106	14,463	1,533,078	0.1	1.0
Total		**12,789,720**	**155,776,218**	100	100
Actual catch			**155,301,000**		
Residual (%)			**0.3**		

Ten years on from the first implementation of the PRPA, Fig. 7.11 shows the cohorts the measure was aimed to preserve in grey. The presence of 9 year old fishes is noteworthy, as this age corresponds exactly to the ABFT born in the year the PRPA was adopted (2007). Moreover, unlike previous decades the presence of large fishes is highly significant, something which ties in with the latest reports of the assessment group, in which a great increase was reported in the biomass of the spawning stock (ICCAT 2012, 2014, 2017).

Based on an analysis of the periods 1966–1986 (Cort 1990) and 1966–1996 (Cort and Ortiz de Zárate 1997), Table 7.4 presents a summary of catches converted to ages of the Bay of Biscay fishery extended to 62 years (1949–2010).

The results confirm something of great importance, which is that 97.5% of the catch (in number of fishes) of this fishery is made up of juveniles (<5 years, Fig. 7.12),

Fig. 7.12 Juvenile specimens (2–3 years; 8–15 kg) caught in the Bay of Biscay (1980s). (Documentary archive, *IEO*)

and it can be confirmed that the exercise was performed correctly since the total catch deriving from the mean weight at age applied to each age is only 0.3% above the real catch in this period (155,301 t). The results of this review were presented to the SCRS in 2016 (Cort 2017).

7.12 Fishing Effort and Abundance Index

The fishing effort of the tuna fleet is monitored by local observers in the largest ports of the north of Spain, both those where only ABFT is landed (Fuenterrabía) and those where landings also include albacore tuna (Guetaria, Ondárroa, Bermeo and Santoña). It is also monitored by the log books kept by some fishing vessels. The unit of fishing effort was *Days at sea*.

Following recommendations made by the SCRS in the 1980s, the Bay of Biscay fishery has been used in assessments as an indicator of the abundance of Atlantic eastern juveniles, and so since then a nominal abundance index has been used for age groups 2 and 3 (fishes of 7–30 kg), i.e. an indicator that included catches of this age group and the days at sea used to catch them (Cort 1995; Ortiz de Zárate and Rodríguez-Cabello 2000). But following the recommendations of the scientific committee this index was standardised because the committee considered that other factors, such as age, the year and month of the catch, the crew number or the number of live bait tanks should be considered in the analysis in order to develop the index more precisely (Rodríguez-Marín et al. 2003; Santiago et al. 2012). As this is a very well defined fishery both in space and time there were no significant variations in results with respect to the nominal index. Age groups 2 and 3 (15–25 kg) were selected for the standardized index.

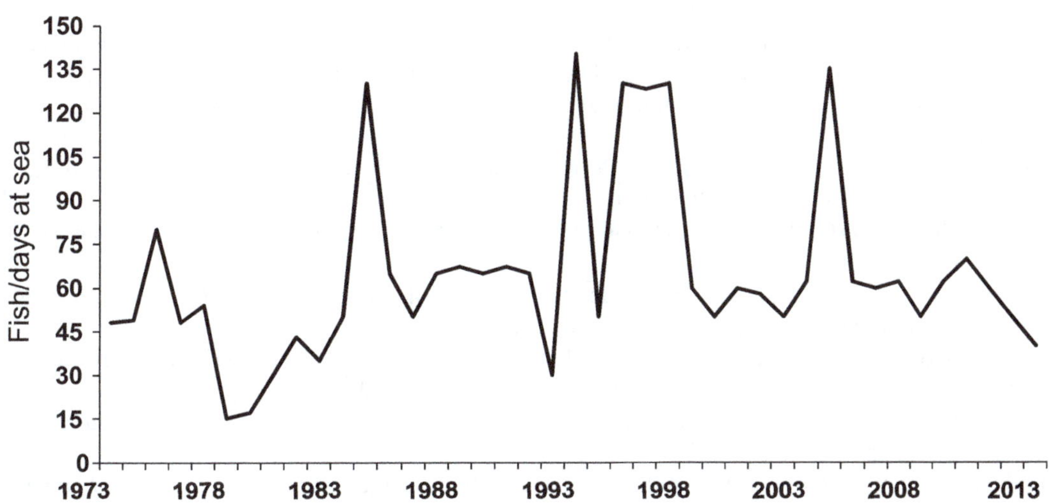

Fig. 7.13 Abundance index (expressed in fishes/days at sea) of bluefin tuna juveniles (ages 2–3) in the Bay of Biscay

For better interpretation, Fig. 7.13 combines the results of the nominal and the standardised indices. The abundance index (expressed in fishes/days at sea) of juvenile bluefin tuna in the Bay of Biscay remains stable throughout the series studied, though there are evident fluctuations and maximums that may be a reflection of years of good recruitments. The fall in recent years was caused by changes in fishing strategy since the implementation of the PRPA in 2008.

The lack of reliability of this index since the PRPA was introduced has made it necessary to seek new ways of obtaining information on the abundance of juveniles in this fishery. To this end, in recent years AZTI has been developing an index based on acoustics obtained in annual surveys (Goñi et al. 2017). In a few years the series may have become large enough to use in assessments.

As we have seen, the ABFT fishery of the Bay of Biscay plays a very important role in the general context of research into the species. In this area field research, samplings and statistical monitoring of the fishery have been carried out for decades and have contributed greatly to our knowledge of the species and improvements in the quality of assessments. All of this work means that the basic tasks of advising administrators, the fishing sector and society as a whole can be performed more effectively.

References

Addis P, Secci M, Pischedda M, Laconcha U, Arrizabalaga H (2014) Geographic variation of body morphology of the Atlantic bluefin tuna (*Thunnus thynnus*, Linnaeus, 1758). J Appl Ichthyol 1–7. ISSN 0175-8659. https://doi.org/10.1111/jai.12453

Aloncle H (1973) Marquage de thons dans le Golfe de Gascogne. Col Vol Sci Pap ICCAT 1:445–458

Aloncle H, Hamre J, Rodríguez-Roda J, Tiews K (1977) Sixth report of the bluefin tuna working group. ICES Coop Res Rep 71:49 pp.

Aloncle H, Hamre J, Rodríguez-Roda J, Tiews K (1981) Seventh report of the bluefin tuna working group. ICES Coop Res Rep 100:70 pp.

Arregui I, Galuardi B, Goñi N, Lam CH, Fraile I, Santiago J, Lutcavage M, Arrizabalaga H (2018) Movements and geographic distribution of juvenile bluefin tuna in the Northeast Atlantic, described through internal and satellite archival tags. ICES J Mar Sci. https://doi.org/10.1093/icesjms/fsy056

Arrizabalaga H, Arregui I, Cosgrove R (2008) Review of Azti-Tecnalia's tuna tagging activities. Col Vol Sci Pap ICCAT 62:2036–2040

Bard FX, Bessineton C, Cendrero O, Dao JC (1973) La pêcherie de thon rouge du Golfe de Gascogne. Col Vol Sci Pap ICCAT 1:397–412

Bard FX, Cort JL (1979) État de la pêcherie franco-espagnole de thon rouge (*Thunnus thynnus*) du golfe de Gascogne. Col Vol Sci Pap ICCAT 8:317–323

Bard FX, Cort JL (1980) Evaluation du recrutement apparent de thon rouge (*Thunnus thynnus*) en Atlantique est à l´ouest de Gibraltar. Col Vol Sci Pap ICCAT 9:553–556

Brêthes JC (1978) Campagne de marquage de jeunes thons rouges au large des côtes du Maroc. Col Vol Sci Pap ICCAT 7:313–317

Brêthes JC (1979) Sur les premiers recuperations des thons rouges marqués en juillet 1977 au large du Maroc. Col Vol Sci Pap ICCAT 8:367–369

Compeán-Jiménez G, Bard FX (1983) Growth increments on dorsal spines of eastern Atlantic bluefin tuna (*Thunnus thynnus* (L.)) and their possible relation to migrations patterns. NOAA, Technical Report NMFS, vol 8, pp 77–86

Cort JL (1976) Datos sobre la biología y pesca del atún rojo (*Thunnus thynnus*, L.) en el golfo de Vizcaya. Campaña de 1975 en el puerto de Fuenterrabía. Col Vol Sci Pap ICCAT 5(2):236–241

Cort JL (1980) Nota sobre la captura de grandes atunes (*Thunnus thynnus*) en el mar Cantábrico. Anuario "*Juan de la Cosa*", Institución Cultural de Cantabria, vol III (1979–1980), pp 230–245

Cort JL (1981) Datos biológicos del atún rojo (*Thunnus thynnus*) en relación con la captura de grandes ejemplares en el mar Cantábrico (norte de España). Bol Inst Esp Oceanog 6(3):111–115

Cort JL (1990) Biología y pesca del atún rojo, *Thunnus thynnus* (L.), del mar Cantábrico. Doctoral thesis. Publicaciones especiales, IEO, vol 4, 272 pp.

Cort JL (1995) Datos de la pesquería de atún rojo del mar Cantábrico. Col Vol Sci Pap ICCAT 44(1):289–292

Cort JL (2006) *El cimarrón del Atlántico Norte y Mediterráneo*. Sexto concurso nacional de Ciencia en Acción. Modalidad, *Medioambiente*. Cort JL (coordinador). Instituto Español de Oceanografía. Depósito legal: SA, 494-2006, 80 pp.

Cort JL (2009) The bluefin tuna (*Thunnus thynnus*) fishery in the Bay of Biscay. Col Vol Sci Pap ICCAT 63:94–102

Cort JL (2017) Review of the catch at age of the Bay of Biscay bluefin tuna fishery (1950–2000). Col Vol Sci Pap ICCAT 73(7):2280–2288

Cort JL, Abascal F, Belda E, Bello G, Deflorio M, de la Serna JM, Estruch V, Godoy D, Velasco M (2010) ABFT tagging manual of the Atlantic-wide research programme for bluefin tuna. ICCAT, 47 pp.

Cort JL, Abaunza P (2015) The fall of tuna traps and collapse of the Atlantic Bluefin Tuna, *Thunnus thynnus* (L.), fisheries of Northern Europe in the 1960s. Rev Fish Sci Aquac 23(4):346–373. http://dx.doi.org/10.1080/23308249.2015.1079166

Cort JL, Arregui I, Estruch V, Deguara S (2014) Validation of the growth equation applicable to the eastern Atlantic bluefin tuna, *Thunnus thynnus* (L.), using L_{max}, tag-recapture and first dorsal spine analysis. Rev Fish Sci Aquac 22(3):239–255. https://doi.org/10.1080/23308249.2014.931173

Cort JL, Bard FX (1980) Descripción de la pesquería de atún rojo en el golfo de Vizcaya. Col Vol Sci Pap ICCAT 11:390–395

Cort JL, Cendrero O (1975) La pesca del atún rojo (*Thunnus thynnus*, L.) en el golfo de Vizcaya (1974). Col Vol Sci Pap ICCAT 4:128–132

Cort JL, de la Serna JM, Velasco M (2012) El peso medio del atún rojo (*Thunnus thynnus*) capturado por las almadrabas del sur de España entre 1914-2010. Col Vol Sci Pap ICCAT 67:231–241

Cort JL, Estruch VD, Santos MN, Di Natale A, Abid N, de la Serna JM (2015) On the variability of the length-weight relationship for Atlantic bluefin tuna, *Thunnus thynnus* (L.). Rev Fish Sci Aquac 23(1):23–38. https://doi.org/10.1080/23308249.2015.1008625

Cort JL, Fernández-Pato C, de Cárdenas E (1976) Observations surla maturation sexualle du ton rouge, *Thunnus thynnus* (L.) du golfe de gascogne. Conseil International pour l´Exploration de la Mer. C.M. 1976/J, vol 11, 4 pp.

Cort JL, Martínez D (2010) Posibles efectos del Plan de Recuperación de atún rojo (*Thunnus thynnus*) en algunas pesquerías españolas. Col Vol Sci Pap ICCAT 65:868–874

Cort JL, Nøttestad L (2007) Fisheries of bluefin tuna (*Thunnus thynnus*) spawners in the Northeast Atlantic. Col Vol Sci Pap ICCAT 60:1328–1344

Cort JL, Ortiz de Zárate V (1997) The bluefin tuna (*Thunnus thynnus*) fishery in the Cantabrian Sea (Northeast Atlantic). In: Hancock DA, Smith DC, Grant A, Beumer JP (eds) Developing and sustaining world fisheries resources: the state of science and management. 2nd World fisheries congress proceedings. CSIRO, Australia, pp 55–66

Cort JL, Rodríguez-Marín E (2009) The bluefin tuna (*Thunnus thynnus*) fishery in the Bay of Biscay. Evolution of 5+ group since 1970. In: ICCAT, world symposium for the study into the stock fluctuation of Northern Bluefin Tunas (*Thunnus thynnus* and *Thunnus orientalis*) Including the Historic Periods. Col Vol Sci Pap ICCAT 63:103–107

Cort JL, Rey JC (1979) Marcado de atunes, *Thunnus thynnus* y *Thunnus alalunga* en el golfo de Vizcaya en el verano de 1978. Col Vol Sci Pap ICCAT 8(2):333–336

Cort JL, Rey JC (1984) Distribución geográfica del atún rojo (*Thunnus thynnus*) juvenil del Atlántico Este, Mediterráneo occidental y Adriático. Col Vol Sci Pap ICCAT 20(2):298–318

Creac'h P (1952) Les aspects économique et biologique de la pêche aux thons dans le fond du Golfe de Gascogne pendant la champagne de 1951. Pêche marit 137–138

De la Tourrase G (1951) La pêche aux thons sur la côte basque française et son evolution récente. Rev Trav Off Pêches Mari Tome 18, fas. 1, 66, 42 pp.

Di Natale A, Lino P, López González JA, Neves dos Santos M, Pagá García A, Piccinetti C, Tensek S (2018) Unusual presence of small bluefin tuna YOY in the Atlantic Ocean and in other areas. Col Vol Sci Pap ICCAT 73(6):3510–3514

Furnestin J, Dardignac J (1962) Le thon rouge du Maroc Atlantique (*Thunnus thynnus* Linné). Trav Inst Pêches Marit 26:381–397

Fraile I, Arrizabalaga H, Rooker R (2014) Origin of Atlantic bluefin tuna (*Thunnus thynnus*) in the Bay of Biscay. ICES J Mar Sci 72(2):625–634. https://doi.org/10.1093/icesjms/fsu156

García-Soto C (2006) Oceanografía del golfo de Vizcaya. In: El cimarrón del Atlántico Norte y Mediterráneo. Sexto concurso nacional de Ciencia en Acción, 2005. Modalidad, Medioambiente. Cort JL (coordinador). Instituto Español de Oceanografía. Depósito legal: SA, 494-2006, 80 pp.

Gil J (2006) Estudios medioambientales, tróficos y comportamiento del atún rojo. In: *El cimarrón del Atlántico Norte y Mediterráneo*. Sexto concurso nacional de Ciencia en Acción, 2005. Modalidad, *Medioambiente*. Cort JL (coordinador). Instituto Español de Oceanografía. Depósito legal: SA, 494-2006, 80 pp.

Goñi N, Fraile I, Arregui I, Santiago J, Boyra G, Irigoien X, Lutcavage M et al (2010) On-going bluefin tuna research in the Bay of Biscay (Northeast Atlantic): The "Hegalabur 2009" Project. Col Vol Sci Pap ICCAT 65(3):755–769

Goñi N, Onandia I, López J, Arregui I, Uranga J, Melvin GD, Boyra G, Arrizabalaga H, Santiago J (2017) Acoustic-based fishery-independent abundance index of juvenile bluefin tunas in the Bay of Biscay. Col Vol Sci Pap ICCAT 73(6):2044–2057

Hamre J (1960) Tuna investigation in Norwegian coastal waters 1954–1958. Ann Biol Cons Int Expl Mer 15:197–211

Hamre J, Tiews K (1963) Second report of the bluefin tuna working group. ICES, C.M. 1963, Scombriform Fish Committee, N. 14, 29 pp.

Heldt H (1943) Études sur le thon, la daurade et les muges. Histoires d´écailles et d´hameçons. Stat. Ocean Salammbô, Tunisia, brochure 1:3–47

ICCAT (2008) Report for biennial period, 2006–07. Part II (2007), vol. 1, 276 pp.

ICCAT (2012) Report of the 2012 Atlantic bluefin tuna stock assessment session, Madrid, Spain, 4–11 Sept 2012, 124 pp.

ICCAT (2014) Report of 2014 Atlantic bluefin tuna stock assessment session. Madrid, Spain, 20–28 July 2014, 178 pp.

ICCAT (2017) Report of the 2017 ICCAT bluefin stock assessment meeting. Madrid, Spain, 22–27 July 2017, 106 pp. http://iccat.int/Documents/Meetings/Docs/2017_BFT_ASS_REP_ENG.pdf

ICCAT Database, version 11/(2017). http://www.iccat.int/en/t1.asp

Lamboeuf M (1975) Contribution a la connaissance des migrations des jeunes thons rouges a partir du Maroc. Col Vol Sci Pap ICCAT 4:141–144

Logan JM, Rodríguez-Marín E, Goñi N, Barreiro S, Arrizabalaga H, Golet WJ, Lutcavage M (2010) Diet of young Atlantic bluefin tuna (*Thunnus thynnus*) in eastern and western foraging grounds. Mar Biol 12 pp. https://doi.org/10.1007/s00227-010-1543-0

Le Gall J (1950) Le thon rouge (*Thunnus thynnus* L.) dans les parages de Saint Jean de Luz, en 1949. Ann Biol Cons Int Expl Mer 6:71–72

Le Gall J (1951) Le thon rouge (*Thunnus thynnus* L.) dans le Golfe de Gascogne en 1951. Ann Biol Cons Int Expl Mer 8:82–83

Logan JM, Rodríguez-Marín E, Goñi N, Barreiro S, Arrizabalaga H, Golet WJ, Lutcavage M (2010) Diet of young Atlantic bluefin tuna (*Thunnus thynnus*) in eastern and western foraging ground. Mar Biol, 12 pp. https://doi.org/10.1007/s00227-010-1543-0

Mather III FJ, Bartlett MR, Beckett JS (1967) Transatlantic migrations of young bluefin tuna. J Fish Res Bd Canada 24(9):1991–1997

Medina A, Cort JL, Aranda G, Varela JL, Aragón L, Abascal F (2011) Summary of bluefin tuna tagging activities carried out between 2009 and 2010 in the East Atlantic and Mediterranean. Col Vol Sci Pap ICCAT 66(2):874–882

Merino JM (1997) La pesca. Servicio Central de Publicaciones del Gobierno Vasco, 1167 pp.

Navaz JM (1950) Le thon de la côte basque. Ann Biol Cons Int Expl Mer 6:71–72

Ortiz de Zárate V, Cort JL (1986) Stomach contents study of immature bluefin tuna in the Bay of Biscay. ICES, C.M. 1986/H26, 10 pp.

Ortiz de Zárate V, Rodríguez-Cabello C (2000) Bluefin tuna (*Thunnus thynnus*) bait boat fishery statistics in the Cantabrian Sea in 1998. Col Vol Sci Pap ICCAT 51:872–880

Pommereau G (1955) Évolution de la pêche au thon. Pêche marit. 34(933):605–609

Rey JC, Cort JL (1984) Una clave talla-edad por lectura de espinas para el atún rojo (*Thunnus thynnus*, L.) del Atlantico Este. Col Vol Sci Pap ICCAT 20(2):337–340

Rodríguez-Marín E, Arrizabalaga H, Ortiz M, Rodríguez-Cabello C, Moreno G, Kell LT (2003) Standardization of bluefin tuna, *Thunnus thynnus*, catch per unit effort in the baitboat fishery of the Bay of Biscay (Eastern Atlantic). ICES J Mar Sci 60:1215–1230

Rodríguez-Marín E, Rodríguez-Cabello C, de La Serna JM, Alot E, Cort JL, Ortiz de Urbina JM, Quintans M (2008a) Bluefin tuna (*Thunnus thynnus*) conventional tagging carried out by the Spanish Institute of Oceanography (IEO) in 2005 and 2006. Results and analysis including previous tagging activities. Col Vol Sci Pap ICCAT 62:1182–1197

Rodríguez-Marín E, Barreiro S, Montero F, Carbonell E (2008b) Looking for skin and gill parasites as biological tags for Atlantic Bluefin tuna (*Thunnus thynnus*). Aquat Living Resour 21:365–371

Rooker J, Arrizabalaga H, Fraile I, Secor DH, Dettman DL, Abid N, Addis P, Deguara S, Karakulak FS, Kimoto A, Sakai O, Macias D, Santos MN (2014) Crossing the line: migratory and homing behaviours of Atlantic bluefin tuna. Mar Ecol Prog Ser 504:265–276

Santiago J, Arrizabalaga H, Ortiz M (2012) Standardized CPUE index of the Bay of Biscay baitboat fishery (1952–2011). In: ICCAT-SCRS/2012/100

Tangen M (2009) The Norwegian fishery for Atlantic bluefin tuna. Col Vol Sci Pap ICCAT 63:79–93

Tensek S, Di Natale A, Pagá García A (2017) ICCAT GBYP psat tagging: the first five years. Col Vol Sci Pap ICCAT 73(6):2058–2073

Tensek S, Pagá García A, Di Natale A (2018) ICCAT GBYP tagging activities in phase 6. Col Vol Sci Pap ICCAT 74(6):2861–2872

Tiews K (1978) On the disappearance of bluefin tuna in the North Sea and its ecological implications for herring and mackerel. Rapp. P-v. Reun. Cons. Cons Int Expl Mer 172:301–309

Von Bertalanffy L (1938) A quantitative theory of organic growth (inquiries on growth laws. II). Hum Biol 10:181–213

Chapter 8
A Publication that Sheds Light on the Disappearance of the Eastern Atlantic Bluefin Tuna Spawner in the 1960s

Based on an article recently published, the influence the massive catch of juvenile fishes on the eastern Atlantic spawning stock (Bay of Biscay and Morocco) between 1949 and 2006 is described, the event which probably led to the decline of the traps of the Strait of Gibraltar and the collapse of the northern European fisheries from the 1960s. The results of the population analysis carried out reveal that during the period from 1949 to 1962, according to the most optimistic scenario, the quantity of fishes reaching the spawning phase would not surpass 16%; during the period 1970–2006, this figure would rise to 41%; and in the present due to the prohibition of catching juveniles in most of the fisheries as a result of the PRPA, all of the mortality in the juvenile phase is due to natural causes.

Up until now we have seen the contribution of the ABFT Bay of Biscay fishery to the eastern stock from 1949 to the present highlighting that there is a before and an after of the PRPA. We shall proceed to look at how this fishery has influenced the remaining eastern Atlantic fisheries.

Among the most relevant events of the last decades, in the introduction we cited the overfishing of the 1960s, which brought with it the downfall of the Atlantic traps and the collapse of the northern European fisheries, among them the Norwegian purse seine fishery. These facts have been the subject of numerous studies and publications (Tiews 1978; Fromentin 2002, 2009; Ravier and Fromentin 2001; Nøttestad and Graham 2004; MacKenzie and Myers 2007; Fromentin and Restrepo 2009; Fromentin et al. 2014; Fromentin and Lopuszanski 2013; Bennema 2018), the organization of an ad hoc symposium in 2008 (ICCAT 2009), and diverse hypotheses relating the event to environmental factors and changes in ABFT migratory patterns. Nevertheless, no conclusion was reached that could explain the true reasons behind the "1963 enigma", so-called because it was from that year that the fall of spawner fisheries took place. In a recent publication Faillettaz et al. (2019) attribute these facts to climate variability.

To provide another response, in 2015 the article "*The fall of the tuna traps and the collapse of the Atlantic bluefin tuna, Thunnus thynnus (L.), fisheries of northern Europe from the 1960s*" was published in *Reviews in Fisheries Science & Aquaculture* by Cort and Abaunza (2015). The study contains an extensive review of the catch data of the adult and juvenile Atlantic fisheries between 1914 and 2010 as well as a

J. L. Cort and P. Abaunza, *The Bluefin Tuna Fishery in the Bay of Biscay*, SpringerBriefs in Biology, https://doi.org/10.1007/978-3-030-11545-6_8

review of studies on the biology and fishing of the species with the aim of studying the consequences of juvenile mortality on the adult population of the eastern Atlantic. The article describes how the ABFT juvenile fisheries (<40 kg, <5 years of age) began to proliferate in the eastern Atlantic at the end of the 1940s, the Bay of Biscay being the first place it did so (with the bait boat fishery) and later, in the middle of the 1950s the Atlantic coasts of Morocco (with purse seine). In both cases these were artisanal fisheries, though this did not make it any less significant as we shall now see. In the western Atlantic ABFT juveniles also began to appear in the purse seine catch from the 1960s, however there is ample scientific information available to dismiss any relation between this circumstance and the events that took place in the eastern Atlantic.

According to this study, the author's working hypothesis was to find out if the level of juvenile catches in the Northeast Atlantic in the period 1949–1962 can plausibly explain the disappearance of adult ABFT from the fisheries in northern Europe and the decrease in the yields of the traps of the Strait of Gibraltar and the adult fishery in the Bay of Biscay. The method of analysis was to use a simple deterministic model of cohort analysis in equilibrium based on the principle of parsimony or Occam's razor (Shiflet and Shiflet 2006). Taking into account the available data and the proposed objective, this simple model can provide a useful answer to the hypothesis presented while deliberately disregarding other factors and approaches that would complicate the model without contributing significant improvements to the main results.

To perform the simulation, the classical equations applied in cohort analysis and catch curve analysis were used (Pitcher and Hart 1982; Hilborn and Walters 1992). Only having catch data and an estimate of natural mortality, the model serves to estimate fishing mortality and abundance at age in the stock (Pitcher and Hart 1982). The three main assumptions of the model were: (a) the population has a constant recruitment at age 1 in each of the periods analyzed: 1949–1962, 1970–2006 and 2009 (a necessary condition for the analysis of cohorts in equilibrium); (b) natural mortality is constant = 0.14; (c) migratory movements were not considered in relation to the western stock of ABFT.

In the data specifications, the authors took into account the ICCAT publications regarding the assessment of the ABFT stock (ICCAT 2014b). Following the concept of equilibrium, they considered a constant recruitment in each period of time (although the cohorts of 1974 and 1994 did not fit well with the mean value of the second period). In this way, the decrease in the abundance of the population reflects the decrease in each of the cohorts (Haddon 2011). The following equation was used to calculate fishing mortality at age i (F_i), with an initial population at age $i = 1$, natural mortality of M, and catch at age $i = C_i$ (the catch data are known up to 4 years old):

$$C_i = N_i \frac{F_i}{F_i + M}\left[1 - e^{-(Fi+M)}\right]$$

The software used to solve the equation in F_i was the program Matlab©R2010a.

Table 8.1 Bluefin tuna juveniles (ages 1–4) caught (number of fish) in the fisheries of the Bay of Biscay and Morocco (Atlantic) between 1949–2009. Taken from Cort and Abaunza (2015) (Courtesy of *Taylor and Francis Group*)

Period 1949–1962/Age	Catch (fish)	Period 1970–2006/Age	Catch (fish)	Current period (2009)/Age	Catch (fish)
1	306,982	1	178,662	1	0
2	128,672	2	74,895	2	21,523
3	42,947	3	24,870	3	16,760
4	12,824	4	7,432	4	4,982
Total	491,426	Total	285,859	Total	43,265

The number of fish present in a population will depend upon the initial number of the population and of those that die. In this way, once F_i was obtained, to get the abundance of that age in the following year, the authors applied the survival equation:

$$N_{i+1} = N_i e^{-(Fi+M)}$$

Thus, the different F at ages 1, 2, 3 and 4 were calculated in the three periods considered (1949–1962, 1970–2006 and 2009) as well as the corresponding survivors in the following year (N_2, N_3, N_4 and N_5). The abundance at age 1 $= N_1$ comes from the initial starting assumption for recruitment in each period.

The study contains 8 basic criteria, of which we will cite the most significant for the purposes of understanding the results of the analysis: According to the information coming from the ABFT assessments (ICCAT 2012, 2014), the abundance and potential of this resource in the Mediterranean Sea is much higher than in the Eastern Atlantic. It is logical to think that the number of recruits that leave the Mediterranean Sea for the Atlantic Ocean is less than those remaining in the Mediterranean Sea. Under this assumption, the authors selected three scenarios for the percentage of ABFT recruits that migrate from the Mediterranean to the Atlantic: 20, 30 and 40%. The population values of the initial abundance in the Eastern Atlantic will correspond to each of the percentages.

In Table 8.1 the values used for the analysis during the three periods described are shown. In the most recent period (2009), the effects of ICCAT's recovery plan were already noticeable, as most of the juvenile catch had disappeared.

The study continues with the description of the catch statistics (in weight and in numbers) of ABFT in the eastern Atlantic, (Bay of Biscay and the Atlantic coast of Morocco, fundamentally) for the three periods considered: 1949–1962, 1970–2006 and 2009. In the first period the catch was 83,448 t in weight and 6,879,967 fishes in number (491,426 fishes caught/year). In the second period the catch was 101,800 t, corresponding 10,576,771 fishes (285,589 fishes caught/year) and for the third period the catch was 728.8 t, equivalent to a total of 43,265 fish caught/year.

In Table 8.2 the results of the simulation analyses are shown. Logically, it can be observed how the initial population is a factor of great influence in the estimations of the fishing mortality at age and of the survivors at age assuming, as has previously been stated, that M is constant.

In this study for the period 1949–1962, a recruitment of 2,000,000 ABFT of 1 year of age to the eastern stock is established. Considering the three scenarios of emigration of juveniles from the Mediterranean to the Atlantic of 20, 30 and 40%, it is observed that:

- In the first scenario (20%; 400,000 emigrants), the total mortality exceeds the number of fish recruited, so this result is not possible.
- In the cases of 30 and 40% of emigrants, fishing mortality far exceeds natural mortality in all ages except in age 4 of the 40% emigration rate scenario. The consequence of this total mortality is a survival curve with a very pronounced decrease, showing very low survival rates of 3% and 16% at the end of the four years considered in the second and third scenarios respectively (see Table 8.2; Fig. 8.1).

The recruitment to age 1 in the eastern stock of ABFT for the period 1970–2006 was assumed at 3,000,000 fish. Considering again the three scenarios of juvenile emigration (20, 30 and 40%), this study showed that:

- *F* at ages 1 and 2 is still higher than natural mortality, especially in the 20 and 30% emigration scenarios, although less pronounced than in the 1949–1962 period. In the same way, the resulting total mortality generates survival curves with less pronounced decreases than in the previous case, resulting in survival rates at the end of the four years of 26 and 41% (Table 8.2, Fig. 8.1).

For the third period (year 2009), the estimated recruitment of ABFT to age 1 is around 4,000,000 fish. The analysis of the three emigration scenarios of juveniles from the Mediterranean to the Atlantic of 20, 30 and 40% shows that:

- The population dynamics of ABFT in the Northeast Atlantic for the juvenile age groups is determined by the natural mortality rate, since fishing mortality is practically nil. As a result, the survival curves show much gentler decreases and significantly higher survival rates at the end of the four years (53 and 55%) than in the previous periods (Table 8.2, Fig. 8.1).

Based on the results presented above, the study concludes that the high values of fishing mortality in the juvenile ages of ABFT (well above natural mortality) during the period 1949–1962 and also in ages 1 and 2 during the period 1970–2006 demonstrate the great importance of fishing activity on a juvenile population for the future of sustained spawner fisheries in the eastern Atlantic.

It should be mentioned that by applying the natural mortality values of the last review carried out by the ABFT assessment group (ICCAT 2017), which point to values much higher than those used in the present study, the result of the analysis would have been much more pessimistic.

Table 8.2 Result of the analysis of the juvenile population. The recruitment values were taken from the analysis by the bluefin tuna assessment group of the SCRS (ICCAT, 2014) F = fishing mortality; Z (total mortality) = $F + N$ (natural mortality). Taken from Cort and Abaunza (2015) (Courtesy of *Taylor and Francis Group*)

Period	Scenario 1	20% outgoing	Final population	Mortality Z (%)	Survival (%)	F_1	F_2	F_3	F_4
	Recruitment	Initial population							
1949–1962	2,000,000	400,000	0	100	0	1.7	–	–	–
1970–2006	3,000,000	600,000	154,070	74	26	0.40	0.30	0.10	0.04
2009	4,000,000	800,000	423,609	47	53	0.0	0.03	0.03	0.01
Period	Scenario 2	30% outgoing	Final population	Mortality Z (%)	Survival (%)	F_1	F_2	F_3	F_4
	Recruitment	Initial population							
1949–1962	2,000,000	600,000	16,231	97	3	0.80	0.90	0.80	0.60
1970–2006	3,000,000	900,000	325,166	64	36	0.20	0.10	0.06	0.02
2009	4,000,000	1,200,000	652,088	46	54	0.0	0.02	0.02	0.01
Period	Scenario 3	40% outgoing	Final population	Mortality Z (%)	Survival (%)	F_1	F_2	F_3	F_4
	Recruitment	Initial population							
1949–1962	2,000,000	800,000	129,620	84	16	0.50	0.40	0.20	0.10
1970–2006	3,000,000	1,200,000	496,415	59	41	0.20	0.10	0.04	0.01
2009	4,000,000	1,600,000	880,570	45	55	0.0	0.02	0.02	0.01

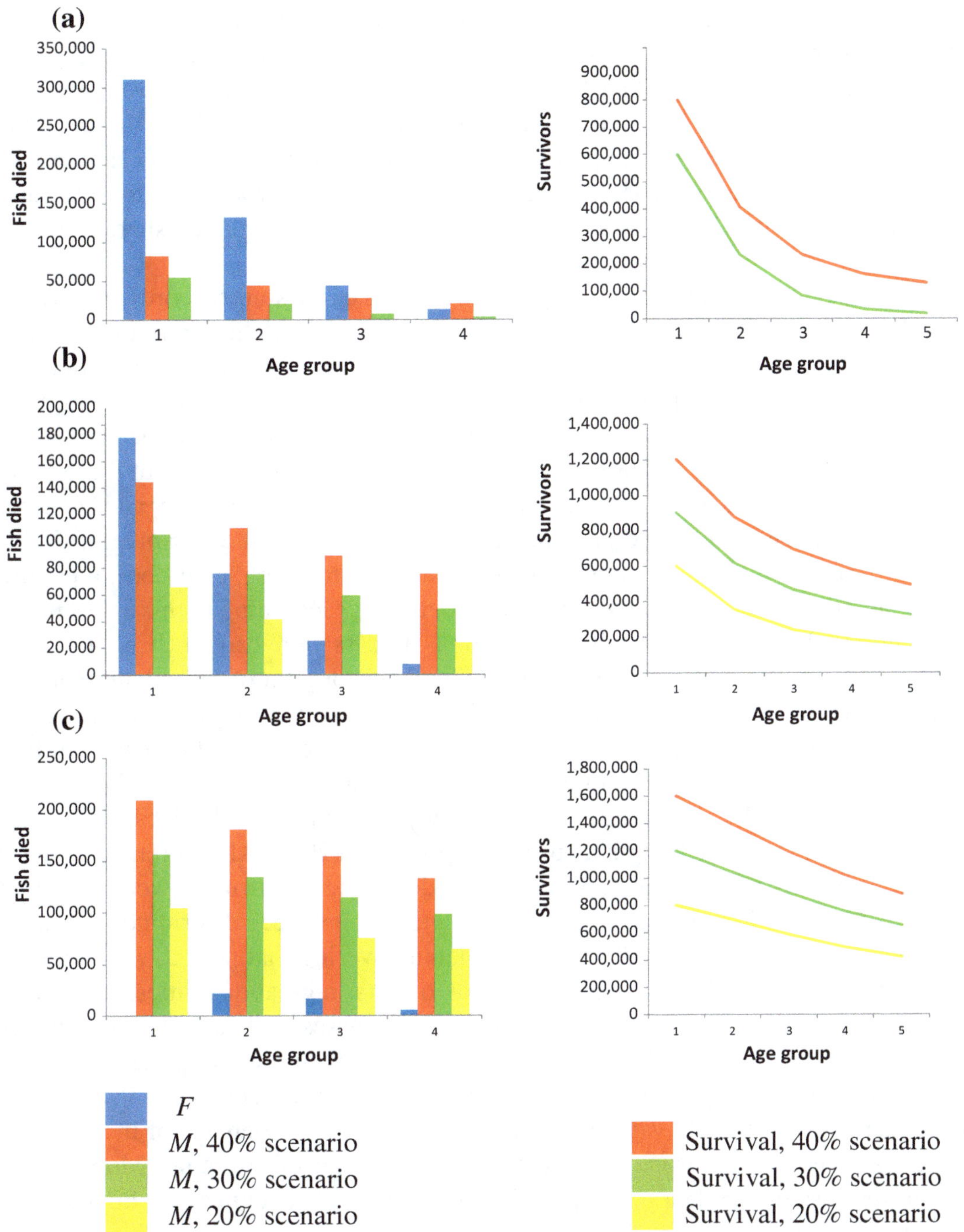

Fig. 8.1 Left column: comparison of the number of fishes/age (ages 1 to 4) dying due to fishing mortality (*F*) and due to natural mortality (*M*) in each of the scenarios and in each of the periods. Right column: survival curves/age in each of the scenarios and in each of the periods. Above (**a**): period 1949–1962; Central (**b**): period 1970–2006; Below (**c**): present period: 2009. (Adapted from Cort and Abaunza 2015)

From the results of the analysis, the survival of juveniles in the 1949–1962 period, after having passed through the juvenile fisheries (Morocco and Bay of Biscay) for 4 years and a mean of 491,426 fishes/year having been caught, in the most optimistic scenario the quantity of fishes reaching the spawning phase would not surpass 16% (Table 8.1). In the intermediate period (1970–2006), now with a mean catch of less than the previous period (285,859 fishes/year), this figure would rise to 41%; and now in the present, with the practical disappearance of the juvenile fisheries, all the mortality in the juvenile phase is due to natural causes.

The Fig. 8.2 represents graphically the conclusion of the study according to which it is demonstrated that the juveniles represented by the circle in the figure above it should have been the future spawners of the other two figures. Special attention should be paid to the quantities caught that appear on the y-axis of the three figures.

The article verifies that the assessments of the resources of the eastern stock carried out by the ABFT assessment group of the SCRS do not detect these facts since they consider the stock as a unit (eastern Atlantic + Mediterranean) and because, moreover, the database of the juvenile catches used in the last assessment in 2014, for the period 1949–1962, is highly underestimated providing, as it does, a picture of the situation that is far from the reality. Specifically, according to this assessment, between 1950 and 1962 1,860,000 ABFT of 1–4 years were caught (ICCAT 2014), when in reality the figure was 6,559,000. The incorporation of these latter catches in the model in the following assessment (in 2017) would have greatly changed the results during the initial phase of the fishery giving a more realistic view than that offered by the results of 2014 in addition to bringing a vision of how the state of the fishery was at the beginning of the intensive exploitation in the 1950s (Fig. 8.3; ICCAT 2014). But something unexpected happened in 2017. All references to these facts were eliminated by removing from the model all the data from between 1950 and 1967 (ICCAT 2017). Why were these data deleted? It had been determined that little information (biological sampling) was available covering that period, which meant that substitutions would have had to be made of data in great quantities, something which, although true, is commonly done in the SCRS assessment groups. The authors consider that the deletion of these data from the reports of the group meant the loss of an important part of the history of the eastern Atlantic bluefin tuna fisheries, as well as the loss of hugely valuable information from the golden age of the spawner fisheries of northern Europe and the traps of the Strait of Gibraltar. Moreover, it meant the loss of the data relating to the overfishing of juveniles that affected the spawner fisheries for years afterwards.

We can conclude this section by stating that the origin of the crisis of the ABFT Atlantic spawner fisheries in the 1960s (traps, Strait of Gibraltar and northern Europe) was possibly the result of the intensive fishing of spawners for a period of 5–10 years and of juveniles in the Bay of Biscay (from 1949) and in waters of Morocco (from 1958). The crisis arose when ICCAT had not yet been founded, so nobody can be blamed for what happened. But the fact that international organizations responsible for defending the preservation of fishing resources now exist does not guarantee the fulfillment of the preservation policies that they themselves determine. It is the political will of the states forming part of these organizations that will ultimately lead

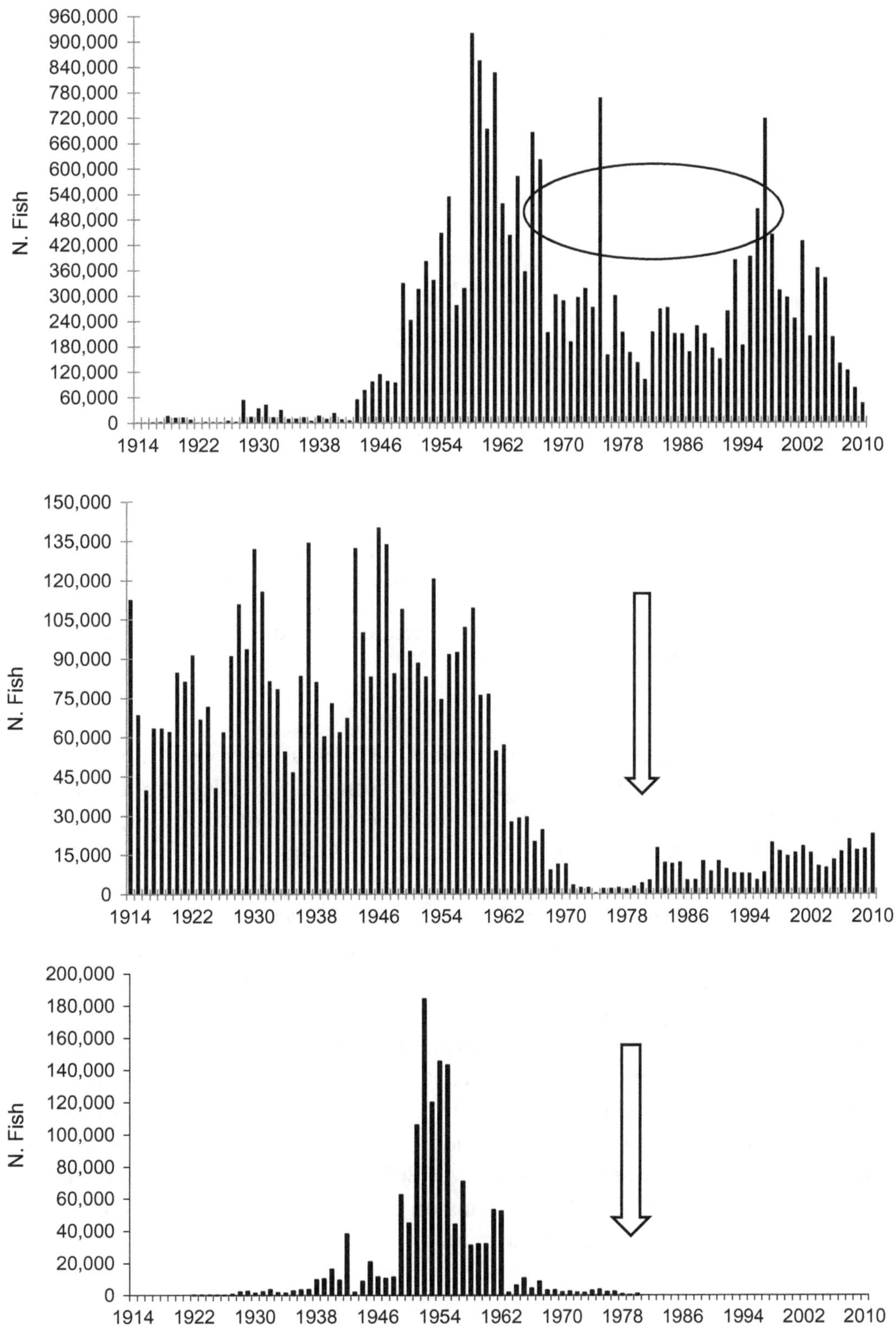

Fig. 8.2 *Above*: juvenile bluefin tuna catches in the fisheries of the Atlantic part of the eastern stock (Morocco and Bay of Biscay). *Central*: bluefin tuna spawner catches in the Atlantic traps of the Strait of Gibraltar (Spain, Morocco and Portugal). *Lower*: bluefin tuna spawner catches in the northern European fisheries

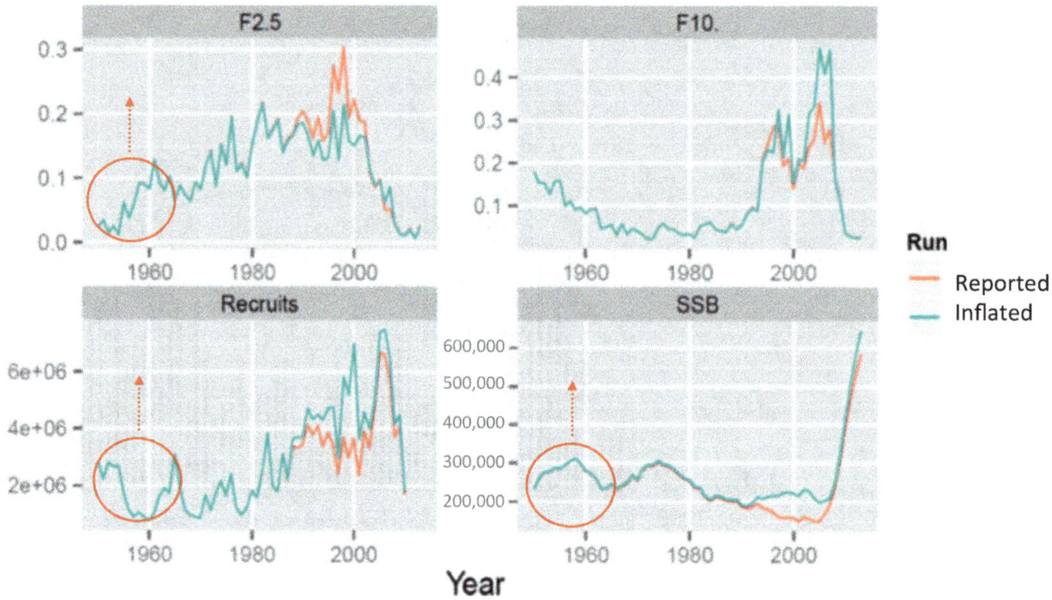

Fig. 8.3 Estimates of fishing mortality (of ages 2 to 5 and 10+) in the upper part; biomass of the spawning stock (in t) and recruitment (in number of fishes) in the lower part according to the population analysis made in 2014. *Red line*: declared catch. *Blue line*: inflated catch. (Courtesy of ICCAT). The red circles/arrows represent the increase expected following application of the analysis of the data supplied by Cort and Abaunza (2015)

to these measures being correctly implemented. ABFT is a clear example of the failure to enforce the measures adopted for its conservation even decades after ICCAT was established (WWF 2008). Nevertheless, under the pressure of environmental NGOs from the 1990s and the requests of Sweden in 1993 (Safina 1993) and Monaco in 2009 (Fromentin et al. 2014) to include bluefin tuna in Appendix 1 of the *Convention of International Trade in Endangered Species of Wild Fauna and Flora* (*CITES*), the situation has been turned around completely such that now, under the monitoring of ICCAT's Pluriannual Recovery Plan (PARP) whose implementation began in 2008, the resources of this species are apparently in much better shape (ICCAT 2017).

There have been many theories to explain what happened to the bluefin tuna spawner population from 1960 and what it was that made this species disappear from the fisheries of northern Europe. In general, it had been put down to environmental factors, the scarcity of prey (herring and mackerel) or to changes in the migratory behaviour of bluefin tuna (ICCAT 2009). Nevertheless, until now nobody had demonstrated the effects of the expansion of the juvenile fisheries in the eastern Atlantic region (Bay of Biscay and Morocco).

References

Bennema FP (2018) Long-term occurrence of Atlantic bluefin tuna *Thunnus thynnus* in the North Sea: contributions of non-fishery data to population studies. Fish Res 199:177–185. https://doi.org/10.1016/j.fishres.2017.11.019

Cort JL, Abaunza P (2015) The fall of tuna traps and collapse of the Atlantic Bluefin Tuna, *Thunnus thynnus* (L.), fisheries of Northern Europe in the 1960s. Rev Fish Sci Aquac 23(4):346–373. http://dx.doi.org/10.1080/23308249.2015.1079166

Faillettaz R, Beaugrand G, Goberville E, Kirby RR (2019) Atlantic Multidecal Oscillations drive the basin-scale distribution of Atlantic bluefin tuna Sci Adv 5 eaar6993

Fromentin JM (2002) Final Report of STROMBOLI-EU-DG XIV Project 99/022. European Community-DG XIV, Brussels, 109 pp

Fromentin JM (2009) Lessons from the past: investigating historical data from bluefin tuna fisheries. Fish Fish 10:197–216

Fromentin JM, Bonhommeau S, Arrizabalaga H, Kell LT (2014) The spectre of uncertainty in management of exploited fish stocks: The illustrative case of Atlantic bluefin tuna. Mar Policy 47:8–14

Fromentin JM, Lopuszanski D (2013) Migration, residency, and homing of bluefin tuna in the western Mediterranean Sea. ICES J. Mar. Sci. https://doi.org/10.1093/icesjms/fst157

Fromentin JM, Restrepo V (2009) A year-class curve analysis to estimate mortality of Atlantic bluefin tuna caught by the Norwegian fishery from 1956–1979. Collect Vol Sci Pap ICCAT 64(2):480–490

Haddon M (2011) Modelling and quantitative methods in fisheries, 2nd edn. Chapman & Hall/CRC, Boca Raton, 449 pp

Hilborn R, Walters CJ (1992) Quantitative fisheries stock assessment. Choice, dynamics and uncertainty. Chapman & Hall, New York, 570 pp

ICCAT (2009) Report of the world symposium for the study into the stock fluctuation of northern bluefin tunas (*Thunnus thynnus* and *Thunnus orientalis*), including the historical periods. Col Vol Sci Pap ICCAT 63:1–49

ICCAT (2012) Report of the 2012 Atlantic bluefin tuna stock assessement session. Madrid, Spain, 4–11 Sept 2012, 124 pp. http://www.iccat.int/Documents/Meetings/Docs/2012_BFT_ASSESS.pdf

ICCAT (2014) Report of 2014 Atlantic bluefin tuna stock assessment session. Madrid, Spain, 20–28 July 2014, 178 pp

ICCAT (2017) Report of the 2017 ICCAT bluefin stock assessment meeting. Madrid, Spain, 22–27 July 2017, 106 pp. http://iccat.int/Documents/Meetings/Docs/2017_BFT_ASS_REP_ENG.pdf

MacKenzie BR, Myers RA (2007) The development of the northern European fishery for north Atlantic bluefin tuna (*Thunnus thynnus*) during 1900–1950. Fish Res. https://doi.org/10.1016/j.fishres.2007.01.013

Nøttestad L, Graham N (2004) Preliminary overview of the Norwegian fishery and science on Atlantic bluefin tuna (*Thunnus thynnus*). Scientific report from Norway to ICCAT Commission meeting in New Orleans, USA, 15–21 Nov 2004, 12 pp

Pitcher TJ, Hart PJB (1982) Fisheries ecology. Chapman & Hall, London, 414 pp

Ravier C, Fromentin JM (2001) Long-term fluctuations in the eastern Atlantic and Mediterranean bluefin tuna population. ICES J Mar Sci 58:1299–1317

Safina C (1993) Bluefin Tuna in the West Atlantic: negligent management and the making of an endangered species. Conserv Biol 7:229–234. http://www.seaweb.org/resources/articles/writings/safina2.php

Shiflet AB, Shiflet GW (2006) Introduction to computational science. Modelling and simulation for the sciences. Princeton University Press, Princeton and Oxford, 554 pp

Tiews K (1978) On the disappearance of bluefin tuna in the North Sea and its ecological implications for herring and mackerel. Rapp P-v Reun Cons Cons Int Expl Mer 172:301–309

WWF (2008) Race for the last bluefin. WWF Mediterranean Project. Zurich, 126 pp. https://www.wwf.or.jp/activities/lib/pdf/0811med_tuna_overcapacity.pdf

Chapter 9
Epilogue

9.1 Epilogue

The traps that catch bluefin tuna in the Strait of Gibraltar have provided work, wealth and food for thousands of years, and are additionally a bottomless source of information for historians, scientists and the public in general. Fishing for this species both in the Atlantic and the Mediterranean has always been sustainable, no written testimony ever having certified that human intervention had affected the overexploitation of the species. But that all changed at the end of the 1940s following the Second World War when new fisheries began to develop in the Atlantic. The main ones were those of northern Europe, mainly in Norway, with the introduction of purse seine, in which spawners were caught, and in the Bay of Biscay mainly targeting juveniles. In 1958 purse seine fishing for juveniles began in Morocco (the Atlantic part).

When in 1963 the catches of the traps in the Strait of Gibraltar suddenly fell and a few years later the fisheries of northern Europe collapsed, nobody knew what was happening nor why. This crisis in catches brought with it the closure of most of the traps and the dissolution of the Tuna Trap Fishing National Consortium, creating misery in a sector that had survived for millennia.

In the article by Cort and Abaunza (2015) the fundamental factor behind those events is shown to be related to the massive juvenile exploitation that began in the Bay of Biscay in 1949. Thus, the analysis made on the Atlantic ABFT juvenile population reveals a clear interaction between the juvenile fisheries of the Bay of Biscay and Morocco (Atlantic part) and those of spawners in the Strait of Gibraltar and northern Europe. The overfishing of juveniles that took place in these fisheries from the 1950s until the 2000s left a gap in the generations of spawners for decades. The first outstanding effect took place in 1963, the year in which the set of ABFT cohorts that had made up the catches of the traps and northern European fisheries had already passed, in at least one year and in most cases four, through the juvenile fisheries of the eastern Atlantic in which 6.9 million ABFT juveniles had been caught between 1949 and 1962 (491,426 fish/year). Juvenile catches on a similar scale were the main cause of the lower recruitment from juvenile ages to adults, which left future

J. L. Cort and P. Abaunza, *The Bluefin Tuna Fishery in the Bay of Biscay*,
SpringerBriefs in Biology, https://doi.org/10.1007/978-3-030-11545-6_9

generations of spawners very much reduced. This led to the immediate decline of the spawner fisheries which brought with it the collapse of the northern European fisheries at the beginning of the 1980s. The traps, however, were able to survive albeit at CPUE levels three times lower than those obtained between 1914 and 1950 (Cort et al. 2012).

The analyses made for the periods 1970–2006 and the present (2009) shows that the catches prior to the implementation of the PARP in 2008 (285,859 fish/year) were not sustainable, which meant very high fishing mortality rates (F) that prevented the recruitment of fishes to the spawner fisheries. With the reduction of the juvenile catch, which is now ongoing, it has all returned to how it was historically. The abundance of spawners in our times is a global phenomenon in the fisheries of the eastern stock (ICCAT 2017).

With the publication of these results the authors have contributed to the clarification of facts that occurred over 55 years ago and which led to the fall of two of the oldest fisheries of our seas. This information was supplied in 2016 to the ABFT assessment group of the SCRS, which should have paid it the attention that a find of this nature deserved and taken note in order to, at the very least, avoid a repetition of any similar cases in the future. Nevertheless, the group failed to make any reference to these facts in its latest analysis and report in 2017 and so their origins and terrible consequences have been buried.

We wish to conclude with the phrase of G. Santayana (1863–1952), who said: "Those who cannot remember the past are condemned to repeat it" (Santayana 1905).

References

Cort JL, Abaunza P (2015) The fall of tuna traps and collapse of the Atlantic Bluefin Tuna, *Thunnus thynnus* (L.), fisheries of Northern Europe in the 1960s. Rev Fish Sci Aquac 23(4):346–373. http://dx.doi.org/10.1080/23308249.2015.1079166

Cort JL, de la Serna JM, Velasco M (2012) El peso medio del atún rojo (*Thunnus thynnus*) capturado por las almadrabas del sur de España entre 1914–2010. Col Vol Sci Pap ICCAT 67:231–241

ICCAT (2017) Report of the 2017 ICCAT bluefin stock assessment meeting. Madrid, Spain, 22–27 July 2017, 106 pp. http://iccat.int/Documents/Meetings/Docs/2017_BFT_ASS_REP_ENG.pdf

Santayana G (1905) The life of reason. Reeditado en 2005 por Editorial Tecnos. ISBN 9788430942510, 320 pp

www.ingramcontent.com/pod-product-compliance
Lightning Source LLC
LaVergne TN
LVHW082238200926
848484LV00008B/2372

9781013272851